Stewart's Calculus

Second Edition

Volume I
Study Guide
Early Transcendentals

Richard St. Andre
Central Michigan University

Brooks/Cole Publishing Company
Pacific Grove, California

Brooks/Cole Publishing Company
A Division of Wadsworth, Inc.

Printed in the United States of America

10 9 8 7 6 5 4 3 2

ISBN 0-534-13831-4

Cartoons courtesy of Sidney Harris. Used by permission.

Contents

Contents

Contents

Please cut page down center line.
Use the half page to cover the right
column while you work in the left.

Review and Preview

"WE HAVE REASON TO BELIEVE BINGLEMAN IS AN IRRATIONAL NUMBER HIMSELF."

Numbers, Inequalities, and Absolute Values

Concepts to Master

A. Real Numbers; Intervals
B. Solving inequalities
C. Absolute value; Triangle Inequality

Summary and Focus Questions

A. The set of <u>real numbers</u>, denoted $\mathcal{R}$, contains both rationals $\left(\text{expressible as } \dfrac{m}{n}\right.$

where m and n are integers with $n \neq 0$, and expressible as repeating decimals) and irrationals (nonrepeating decimals).

The set of all numbers between a pair of real numbers a and b is called an <u>interval</u>. The types of intervals include:

open interval $\quad (a, b) = \{x \mid a < x < b\}$

closed interval $\quad [a, b] = \{x \mid a \leq x \leq b\}$

and infinite intervals:

open ray $(a, \infty) = \{x \mid a < x\}$

open ray $(-\infty, b) = \{x \mid x < b\}$

closed ray $[a, \infty) = \{x \mid a \leq x\}$.

1) True or False:

Every real number is either rational or irrational.

True

2) Between every two distinct real numbers there is another real.

True, if a and b are reals then $\dfrac{a+b}{2}$

is another real number that is halfway between a and b.

3) True or False:

$\dfrac{\pi}{6}$ is rational.

False.

4) True or False:

3.3 is an element of [3.1, 3.6).

True.

5) True or False:

4.8 is an element of (4.8, 5.8).

False.

B. How to solve an inequality with a single variable x will depend on the initial form of the inequality. A linear inequality (such as $4x + 7 < 21$) is solved by reducing one side to just x. Most other inequalities involve getting 0 on one side of the inequality, then factoring the other side.

6) Solve $4x + 7 < 21$.

$4x + 7 < 21$
Add -7 to each side:
$4x + 7 - 7 < 21 - 7$
$4x < 14$
Divide each side by 4:
$x < \dfrac{14}{4}$ or $x < \dfrac{7}{2}$.

The solution is the interval $\left(-\infty, \dfrac{7}{2}\right)$.

7) Solve $\dfrac{x^2 - 5x}{x + 2} \leq 0$.

We already have 0 on one side.

We factor: $\dfrac{x\,(x - 5)}{x + 2} \leq 0$.

At $-2, 0, 5$ the factors change signs.

I	II	III	IV

$$\xleftarrow{\qquad} \underset{-2}{+} \qquad \underset{0}{+} \qquad\qquad \underset{5}{+} \xrightarrow{\qquad}$$

I	II	III	IV
$x + 2 < 0$	$x + 2 > 0$	$x + 2 > 0$	$x + 2 > 0$
$x < 0$	$x < 0$	$x > 0$	$x > 0$
$x - 5 < 0$	$x - 5 < 0$	$x - 5 < 0$	$x - 5 > 0$

The inequality holds in regions I and III.
Note that $x = -2$ is not a solution but
$x = 0$, $x = 5$ are. The set of solutions is
$(-\infty, -2) \cup [0, 5]$.

C. For $x \geq 0$, $|x| = x$; for $x < 0$, $|x| = -x$. You should think of $|x|$ as the distance from x to the origin and $|a - b|$ as the distance between numbers a and b. The following are usually used to solve problems involving absolute values (here a is nonnegative):

$|x| = a$ means $x = a$ or $x = -a$
$|x| < a$ means $-a < x < a$
$|x| > a$ means $x < -a$ or $x > a$

The Triangle Inequality is $|a + b| \leq |a| + |b|$ for all reals a and b.

8) Sometimes, Always, or Never:
$$|x - 3| = x - 3.$$

Sometimes. This is true when $x - 3 \geq 0$, that is, when $x \geq 3$.

9) Solve $|x - 8| \geq 3$.

The distance between x and 8 is at least 3.

Thus $x \geq 11$ or $x \leq 5$.

Using algebra, this inequality is

$x - 8 \geq 3$ or $x - 8 \leq -3$,

thus $x \geq 11$ or $x \leq 5$.

10) Solve $\left| \dfrac{1-x}{x} \right| = 1$.

Either $\dfrac{1-x}{x} = 1$ or $\dfrac{1-x}{x} = -1$.

So $1 - x = x$ or $1 - x = -x$.

Thus $1 = 2x$ or $1 = 0$.

Since $1 = 0$ always fails, $1 = 2x$.

Thus $x = \dfrac{1}{2}$.

11) True or False:

$|a+b| - |a| \leq |b|$.

True. This is the Triangle Inequality with $|a|$ subtracted from both sides.

Coordinate Geometry and Lines

Concepts to Master

A. Distance between two points
B. Slope of a line; Equations of a line
C. Parallel lines; Perpendicular lines
D. Intersection of two lines

Summary and Focus Questions

A. The <u>distance between two points</u> $A(x_1, y_1)$ and $B(x_2, y_2)$ in the Cartesian plane is

$$|AB| = \sqrt{(x_2 - x_1)^2 + (y_2 - y_1)^2}.$$

1) Find the distance between $(5, -7)$ and $(2, 3)$.

Choose A as $(5, -7)$ and B as $(2, 3)$. (The choice is immaterial.)

$$|AB| = \sqrt{(5 - 2)^2 + (-7 - 3)^2}$$
$$= \sqrt{3^2 + (-10)^2} = \sqrt{109}.$$

B. The <u>slope</u> of the line through $A(x_1, y_1)$

and $B(x_2, y_2)$ is $m = \dfrac{y_2 - y_1}{x_2 - x_1}$

when $x_1 \neq x_2$. In the case where $x_1 = x_2$, we have a vertical line with an undefined slope and the equation is simply $x = x_1$.

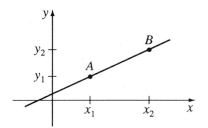

Otherwise the <u>equation of the line</u> through A and B is $y - y_1 = m(x - x_1)$, which may be written in slope-intercept form as:

$$y = mx + b,$$

where $(0, b)$ is where the line intercepts the y-axis.

The general equation of the line has the form

$$Ax + By + C = 0.$$

2) Find the equation of the line through $(4, 7)$ with slope -2.

$$y - 7 = -2(x - 4)$$

3) Is $3x + xy + 4 = 0$ the equation of a line?

<u>No</u>. The coefficients of x and y must be constants. The term xy makes the graph nonlinear.

4) Find the slope of the line through $(17, 6)$ and $(7, -3)$.

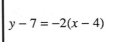

$$m = \frac{6 - (-3)}{17 - 7} = \frac{9}{10}$$

5) Find the equation of the line through $(2, 4)$ and having a y-intercept of 12.

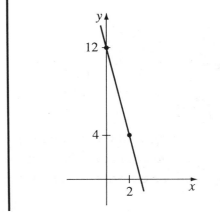

Two points on the line are (2, 4) and (0, 12). The slope is

$$m = \frac{4 - 12}{2 - 0} = \frac{-8}{2} = -4.$$

The equation is $y = -4x + 12$.

C. Lines L_1 and L_2 are parallel if and only if either both are vertical or they have the same slope.

Lines L_1 and L_2 are perpendicular if and only if either one is horizontal and the other is vertical, or if slope of $L_2 = \dfrac{-1}{\text{slope of } L_1}$.

6) Are L_1: $7x - 5y + 3 = 0$
   and L_2: $4x + 6y + 9 = 0$ perpendicular?

No. Rewrite each in slope intercept form.
L_1: $7x - 5y + 3 = 0$

$$-5y = -7x - 3$$
$$y = \frac{7}{5}x + \frac{3}{5}$$

L_2: $4x + 6y + 9 = 0$

$$6y = -4x - 9$$
$$y = -\frac{2}{3}x - \frac{3}{2}$$

The slope of L_2 is $-\frac{2}{3}$ while

$$\frac{-1}{\text{slope of } L_1} \text{ is } \frac{-1}{\frac{7}{5}} = -\frac{5}{7}.$$

7) Is L_1: $y = 7x + 3$ parallel to L_2: $y = 3x + 7$?

No. The slope of L_1 (7) is not the same as the slope of L_2 (3).

D. To determine the intersection of two lines of the form

$$ax + by = c$$
$$rx + sy = t$$

we must "solve two equations simultaneously." Multiply one or both equations by constant(s) so that either x or y has the same coefficient, subtract the equations, and solve for the remaining variable.

1. If one solution exists, the lines intersect at one point.
2. If no solution exists, the lines are parallel and do not intersect.
3. If $0 = 0$ results, the two equations represent the same line.

Here is an example.

Find the intersection of

$$4x + 7y = 15$$
$$6x + 3y = 15.$$

Multiply the first equation by 3 and the second by 2:

$$12x + 21y = 45$$
$$12x + 6y = 30$$

Subtract:

$$15y = 15$$
$$y = 1$$

At $y = 1$, $4x + 7(1) = 15$, $4x = 8$, $x = 2$. The point of intersection is (2, 1).

8) Find the intersection of these pairs of lines.

a) $8x + y = 6$
    $10x + 3y = 4$

Multiply the first equation by 3:

$$24x + 3y = 18$$
$$10x + 3y = 4$$

Subtract: $14x = 14$
$$x = 1$$

At $x = 1$, $8(1) + y = 6$, $y = -2$.

The point of intersection is $(1, -2)$.

b) $2x - 3y = 10$
$4x - 6y = 20$

Multiplying the first equation by 2 yields exactly the second. These are two equations for the <u>same line.</u>

c) $6x + 4y = 20$
$9x + 6y = 40$

Multiply the first equation by 3 and the second by 2.

$$18x + 12y = 60$$
$$18x + 12y = 80$$

Subtracting yields $0 = -20$.

These are parallel lines that <u>do not intersect.</u>

Graphs of Second-Degree Equations

Concepts to Master

A. Graphs of certain circles, parabolas, ellipses, and hyperbolas
B. Shifting a graph horizontally or vertically

Summary and Focus Questions

A. The following summarizes the types of graphs in this section.

Equation	Graph	Description
$(x - h)^2 + (y - k)^2 = r^2$	Circle	Center at (h, k), radius is r.
$y = ax^2 + bx + c$	Parabola	Opens upward if $a > 0$. Opens downward if $a < 0$.
$x = ny^2 + py + q$	Parabola	Opens to the right if $n > 0$. Opens to the left if $n < 0$.
$\dfrac{x^2}{a^2} + \dfrac{y^2}{b^2} = 1$	Ellipse	Center at $(0, 0)$. If $a^2 > b^2$, elongated vertically. If $a^2 < b^2$, elongated horizontally. If $a^2 = b^2$, a circle.
$\dfrac{x^2}{a^2} - \dfrac{y^2}{b^2} = 1$	Hyperbola	Opens upward and downward.
$\dfrac{y^2}{b^2} - \dfrac{x^2}{a^2} = 1$	Hyperbola	Opens to the left and right.
$xy = k$	Hyperbola	Asymptotes along x- and y-axes.

Review & Preview 3: Graphs of Second-Degree Equations

Key concepts to help determine the graph include:

symmetry about the y-axis: the equation is unchanged when x is replaced by $-x$.
symmetry about the x-axis: the equation is unchanged when y is replaced by $-y$.
x-intercept(s): point(s) on the graph where $y = 0$.
y-intercept(s): point(s) on the graph where $x = 0$.

1) Describe the equation form of each:

a)

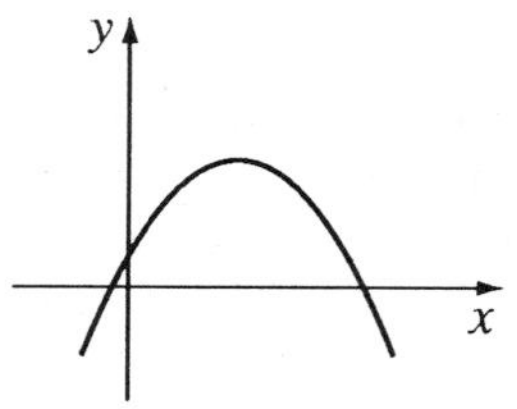

Parabola. $y = ax^2 + bx + c$, with $a < 0$

b)

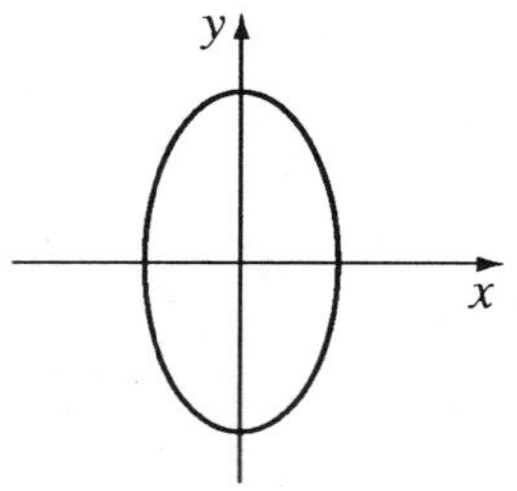

Ellipse. $\dfrac{x^2}{a^2} + \dfrac{y^2}{b^2} = 1$, with $a^2 < b^2$

c)

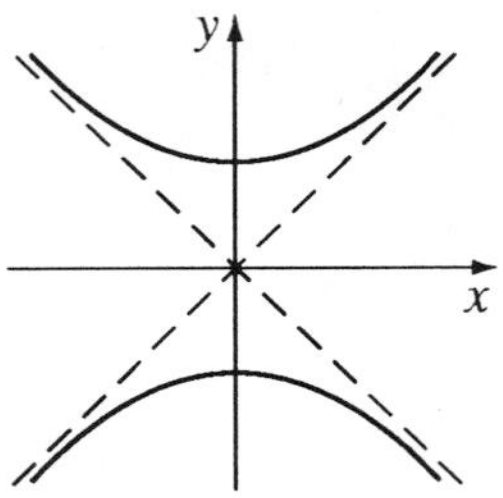

Hyperbola. $\dfrac{y^2}{b^2} - \dfrac{x^2}{a^2} = 1$

2) Sketch the graph of $y = 10 - x^2$.

The graph is a parabola opening downward. $y = 10 - (-x)^2$ is the same equation so the graph is symmetric about the y-axis.
$-y = 10 - x^2$ and $-y = 10 - (-x)^2$ are different equations, so there is no other symmetry.
At $x = 0$, $y = 10$, so $(0, 10)$ is the y-intercept.
$y = 0$ when $10 - x^2 = 0$,
$x^2 = 10$, $x = \sqrt{10}$, $-\sqrt{10}$
The x-intercepts are $\left(\sqrt{10}, 0\right)$, $\left(-\sqrt{10}, 0\right)$.

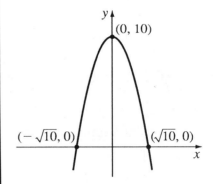

3) What is the equation of the circle in the figure?

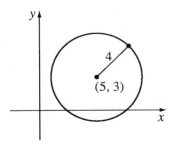

$(x - 5)^2 + (y - 3)^2 = 16$

4) What type of graph is $3x^2 + y^2 - 12x = 13$?

From $3x^2 + y^2$ we see it is an ellipse.
Collect x terms and complete the square.
$3x^2 - 12x + y^2 = 13$
$3(x^2 - 4x \quad) + y^2 = 13$
$3(x^2 - 4x + 4) + y^2 = 13 + 3 \cdot 4$
$3(x - 2)^2 \qquad + y^2 = 27$

$$\frac{(x - 2)^2}{9} + \frac{y^2}{27} = 1$$

B. Let $h > 0$. A graph is <u>shifted h units to the *right*</u> when every occurrence of x in its equation is replaced by $x - h$.

A graph is <u>shifted h units to the *left*</u> when x is replaced by $x + h$.

Let $k > 0$. A graph is <u>shifted k units *upward*</u> when every occurrence of y is replaced by $y - k$.

A graph is <u>shifted k units *downward*</u> when y is replaced by $y + k$.

5) Here is the graph of $y = x^2$:

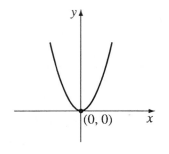

$(0, 0)$

Sketch a graph of each of the following using the given graph:

a) $y = (x + 2)^2$

b) $y - 3 = x^2$

c) $y + 4 = (x - 1)^2$

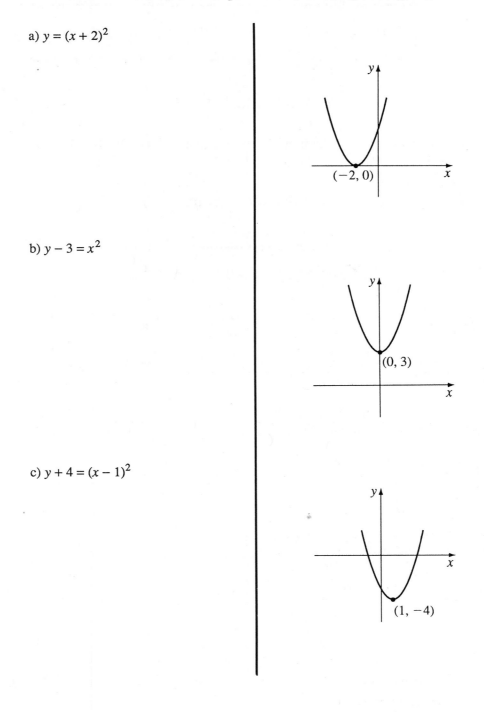

Functions and Their Graphs

Concepts to Master

A. Definition of a function; Domain; Range; Independent variable; Dependent variable
B. Implied domain
C. Graph of a function; Symmetry (even, odd functions)

Summary and Focus Questions

A. A function f is a rule that associates to each number x in a set (the domain) another real number, denoted $f(x)$, the image of x. The set of all images ($f(x)$ values) is the range of f. The variable x is the independent variable and $y = f(x)$ is the dependent variable.

For example, the function S, "to each nonnegative number, associate its square root," has domain $[0, \infty)$ and rule of association given by:

$$S(x) = \sqrt{x}.$$

1) Sometimes, Always, Never:

Both of the points $(2, 5)$ and $(2, 7)$ could be pairs of a function f.

Never. $f(2)$ cannot be both 5 and 7.

2) True or False:

$x = y^2$ defines y as a function of x.

False. Since $(4, 2)$ and $(4, -2)$ satisfy the equation. Note that x is a function of y.

3) The phone company charges a flat rate for local calls of $12.60 per month with an additional charge of $0.05 per call after 55 calls. Express the monthly bill as a function of the number of local calls.

Let x = the number of local calls. Then

$$M(x) = \begin{cases} 12.60, \text{ if } x \in [0, 55] \\ 12.60 + 0.05(x - 55), \text{ if } x \in (55, \infty) \end{cases}$$

B. The domain of a function f, if unspecified, is understood to be the largest set of the real numbers x for which $f(x)$ exists. Thus $f(x) = \dfrac{1}{\sqrt{x - 5}}$ has implied domain $(5, \infty)$. The range of $f(x)$ is $(0, \infty)$.

4) Find the domain of $f(x) = \sqrt{x^2 - 16}$.

For $f(x)$ to exist, $x^2 - 16 \geq 0$.
Thus $(x - 4)(x + 4) \geq 0$. Using
the techniques of Section 1, the solution
set is $x \leq -4$ or $x \geq 4$. The domain is
$(-\infty, -4) \cup [4, \infty)$.

C. The <u>graph</u> of $y = f(x)$ is all points (x, y) in the Cartesian plane that make $y = f(x)$ true. About all we can do now to sketch graphs are plot points and recognize certain graphs (such as lines) from the form of the equation. We can check for some symmetry:

$y = f(x)$ is <u>even</u> means $f(x) = f(-x)$ for all x in the domain.
$y = f(x)$ is <u>odd</u> means $f(x) = -f(-x)$ for all x in the domain.

An *even* function is symmetric about the y-axis.
An *odd* function is symmetric with respect to the origin.

5) Is $f(x) = 10 - 2x$ even or odd?

$f(x) = 10 - 2x$
$f(x) = 10 - 2(-x) = 10 + 2x$
$f(x)$ is <u>not even</u> since $f(x) \neq f(-x)$.
$-f(-x) = -(10 + 2x) = -10 - 2x$
$f(x)$ is <u>not odd</u> since $f(x) \neq -f(-x)$.

6) Sketch the graph of $f(x) = \sqrt{x^2 - 16}$.
(See question 4.)

The function is symmetric about the y-axis because $f(x) = f(-x)$. Here are several computed values:
$(4, 0)$, $(5, 3)$, $(6, \sqrt{20})$, ...
Squaring both sides of $\sqrt{x^2 - 16}$
gives $y^2 = x^2 - 16$, or
$x^2 - y^2 = 16$.
The graph is the top half of a hyperbola.

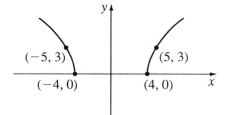

Combinations of Functions

Concepts to Master

A. Sum, Difference, Product, and Quotient of two functions
B. Composition of functions; Writing a function as a composite

Summary and Focus Questions

The purpose of this section is to show how simple functions may be combined to produce more complex combinations. Later, when principles of calculus are developed for simple functions, those principles may be applied to complex functions as well by applying the principles to the complex function's components.

A. Given two functions $f(x)$ and $g(x)$:

1) The <u>sum</u> $f + g$ associates to each x the number $f(x) + g(x)$.
2) The <u>difference</u> $f - g$ associates to each x the number $f(x) - g(x)$.
3) The <u>product</u> fg associates to each x the number $f(x) \cdot g(x)$.
4) The <u>quotient</u> $\dfrac{f}{g}$ associates to each x the number $\dfrac{f(x)}{g(x)}$, provided that $g(x) \neq 0$.

For example, the sum, difference, product, and quotient of $f(x) = x^2$ and $g(x) = 2x - 8$ are:

$$(f + g)(x) = x^2 + 2x - 8$$
$$(f - g)(x) = x^2 - 2x + 8$$
$$fg(x) = x^2(2x - 8)$$
$$\frac{f}{g}(x) = \frac{x^2}{2x - 8}, \text{ for } x \neq 4.$$

The function $h(x) = \dfrac{x^2(x+1)}{2+x}$ may be written as $h = \dfrac{fg}{k}$ where $f(x) = x^2$, $g(x) = x + 1$, and $k(x) = 2 + x$.

1) Find $f + g$ and $\dfrac{f}{g}$ for $f(x) = 8 + x$, $g(x) = \sqrt{x}$.

   What is the domain of each?

$(f + g)(x) = 8 + x + \sqrt{x}$ with domain $= [0, \infty)$.

$\dfrac{f}{g}(x) = \dfrac{8 + x}{\sqrt{x}}$ with domain $= (0, \infty)$.

2) Write $f(x) = \dfrac{2\sqrt{x+1}}{x^2(x+3)}$ as a combination of simpler functions.

There are many answers to this question.

One is $f = \dfrac{hk}{mn}$ where $h(x) = 2$, $k(x) = \sqrt{x+1}$, $m(x) = x^2$, $n(x) = x + 3$.

B. For $f(x)$ and $g(x)$, the <u>composite</u> $f \circ g$ is the function that associates to each x the same number that f associates to $g(x)$, that is:

$$(f \circ g)(x) = f(g(x)).$$

To compute this, first compute $g(x)$, then compute f of that result.

For example, if $f(x) = x^2 + 1$ and $g(x) = 3x$, then for $x = 5$:

$$(f \circ g)(5) = f(g(5)) = f(3(5)) = f(15) = 15^2 + 1 = 226.$$

To determine the components f and g of a composite function, $h(x) = (f \circ g)(x)$, requires an examination of which function is applied first.

For example, for $h(x) = \sqrt{x+5}$, we may use $g(x) = x + 5$ (it is applied first, before the square root) and $f(x) = \sqrt{x}$. Then

$$h(x) = \sqrt{x+5} = \sqrt{g(x)} = f(g(x)).$$

3) Find $f \circ g$ where:

   a) $f(x) = x - 2x^2$ and $g(x) = x + 1$

$$f \circ g(x) = f(g(x)) = f(x + 1)$$
$$= (x + 1) - 2(x + 1)^2.$$

   b) $f(x) = \sin x$ and $g(x) = \sin x$

$$f \circ g(x) = f(\sin x) = \sin(\sin x).$$
This is not the same as $\sin^2(x)$.

4) Rewrite $h(x) = (x^2 + 3x)^{2/3}$ as a composition.

Let $f(x) = x^{2/3}$ and $g(x) = x^2 + 3$.
$$f \circ g(x) = f(g(x))$$
$$= f(x^2 + 3x)$$
$$= (x^2 + 3x)^{2/3}.$$

Types of Functions; Shifting and Scaling

Concepts to Master

A. Constant, power, polynomial, rational, algebraic, and transcendental functions
B. Horizontal and vertical shifting and stretching of functions; Reflecting

Summary and Focus Questions

A. Some of the types of functions that occur in calculus are included in this table.

Type	Description	Examples
Constant	$f(x) = c$	$f(x) = 5$ $f(x) = -\pi$
Power	$f(x) = x^a$ $a = -1, 1, 2, 3, 4,...$	$f(x) = x^7$ $f(x) = x^{-1}$
Root	$f(x) = x^a$ where $a = \frac{1}{2}, \frac{1}{3}, \frac{1}{4}...$	$f(x) = \sqrt[3]{x}$ $f(x) = x^{1/10}$
Polynomial	$f(x) = a_n x^n + \cdots + a_1 x + a_0$ (n is the degree)	$f(x) = 6x^2 + 3x + 1$ $f(x) = -5x^{12} + 7x$
Rational	$f(x) = \dfrac{P(x)}{Q(x)}$,where P, Q are polynomial	$f(x) = \dfrac{7x + 6}{x^2 + 10}$
Algebraic	$f(x)$ is obtained from polynomials, using algebra $(+, -, \cdot, /,$ roots, composition)	$f(x) = \sqrt{16 - x^2}$ $f(x) = \sqrt[3]{\dfrac{x^3}{x+1}}$
Transcendental	Nonalgebraic, including exponential, logarithmic, and trigonometric functions	$f(x) = 2^x$ $f(x) = \log_4(x + 1)$ $f(x) = \cos 3x$

1) True or False:

   $f(x) = x^3 + 6x + x^{-1}$ is a polynomial.

   <u>False.</u>

2) True or False:

   $f(x) = x + \dfrac{1}{x}$ is rational.

   <u>True,</u> $\left(x + \dfrac{1}{x} = \dfrac{x^2 + 1}{x} \right)$

3) True or False:

   Every rational function is an algebraic function.

   <u>True.</u>

4) $f(x) = 6x^{10} + 12x^4 + x^{11}$ has degree ____.

   11.

B. Adding or subtracting a constant to either the dependent or the independent variable shifts the graph up, down, left, or right.

For $c > 0$:

Function	Effect on Graph of $y = f(x)$
$y = f(x - c)$	shift *right* by c units
$y = f(x + c)$	shift *left* by c units
$y = f(x) + c$	shift *upward* by c units
$y = f(x) - c$	shift *downward* by c units

Multiplying either the dependent or the independent variable by a constant will stretch or compress a graph.

For $c > 1$:

Function	Effect on Graph of $y = f(x)$
$y = c\,f(x)$	Stretch *vertically* by a factor of c.
$y = \dfrac{1}{c}\,f(x)$	Compress *vertically* by a factor of c.
$y = f(cx)$	Compress *horizontally* by a factor of c.
$y = f\left(\dfrac{x}{c}\right)$	Stretch *horizontally* by a factor of c.

Changing the sign of either the dependent or independent variable will reflect the graph about an axis.

Function	Effect on Graph of $y = f(x)$
$y = -f(x)$	Reflect about x-axis.
$y = f(-x)$	Reflect about y-axis.

These operations may be combined to produce the graph of a "complex" function from the graph of a simpler function.

5) Given the graph below, sketch each:

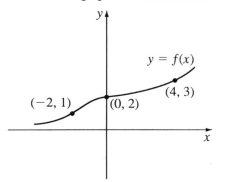

$y = f(x)$

$(4, 3)$

$(-2, 1)$ $(0, 2)$

Review & Preview 6: Types of Functions; Shifting and Scaling

a) $y = 2 f(x)$

b) $y = f(x) - 2$

c) $y = f(-x)$

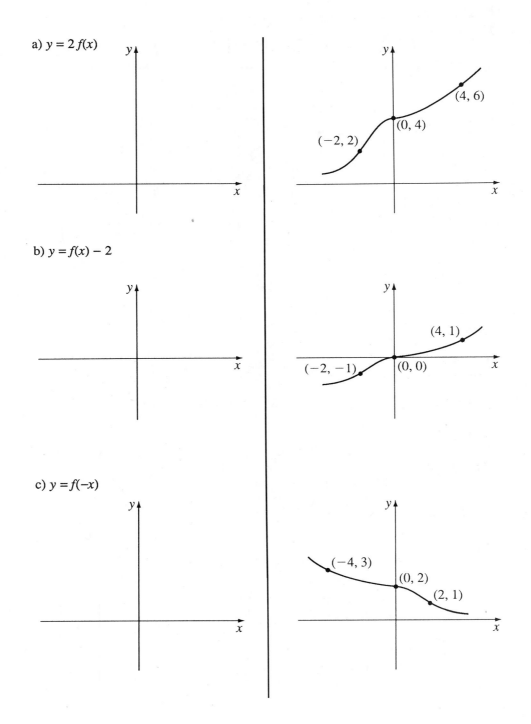

1

Limits and Rates of Change

The Tangent and Velocity Problems

Concepts to Master

A. Slope of secant line; Slope of tangent line
B. Interpretations of slopes as velocities

Summary and Focus Questions

A. A <u>secant line</u> at $P(a, b)$ on the graph of
$f(x)$ is a line joining P and some other point
Q also on $y = f(x)$. To find the slope of the
tangent line at P we select points Q closer
and closer to P. Then the slope of the secant
line becomes a better and better estimate of
the slope of the tangent line at P. Finally our
guess of what value the slopes of the secant
lines approach is what we take as the slope
of the tangent line.

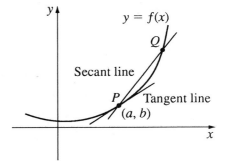

Once we know the slope m of the tangent line, the equation (using P as a point on
the line) is $y - b = m(x - a)$.

1) Let $P(1, 5)$ be a point on the graph of
$f(x) = 6x - x^2$. Let $Q(x, 6x - x^2)$ be on
the graph. Find the slope of the secant line
PQ for each given x value for Q.

x	$f(x)$	Slope of PQ
3		
2		
1.5		
1.01		

x	$f(x)$	Slope of PQ
3	9	$\dfrac{9-5}{3+1} = 2$
2	8	$\dfrac{8-5}{2-1} = 3$
1.5	6.75	$\dfrac{6.75-5}{1.5-1} = 3.5$
1.01	5.0399	$\dfrac{5.0399-5}{1.01-1} = 3.99$

2) Use your answer to question 1 to guess the slope of the tangent to $f(x)$ at P.

The values appear to be approaching 4.

3) What is the equation of the tangent line in question 2?

$y - 5 = 4(x - 1)$.

B. If $f(x)$ is interpreted as the distance along an x-axis of an object from the origin at time x, then

a) the slope of the secant line from P to Q is the average velocity from P to Q.

b) the slope of the tangent line at P is the instantaneous velocity at P.

Thus calculating instantaneous velocities is performed in the same manner as calculating the slopes of tangent lines.

Section 1.1: Tangent and Velocity Problems

4) If a ball is $x^2 + 3x$ feet from the origin at any time x (in seconds), what is the instantaneous velocity when $x = 2$?

At $x = 2$, $f(2) = 2^2 + 3(2) = 10$.

Select x values approaching 2:

x	$f(x)$	Slope of Secant
3	18	$\dfrac{18 - 10}{3 - 2} = 8$
2.5	13.75	$\dfrac{13.75 - 10}{2.5 - 2} = 7.5$
2.1	10.71	$\dfrac{10.71 - 10}{2.1 - 2} = 7.1$

We guess the slope of the tangent, and thus the instantaneous velocity is 7 ft/sec.

5) Let $f(x) = mx + b$, a linear function. Sometimes, Always, or Never:

The average velocity for $f(x)$ from P to Q is the same as the instantaneous velocity at P.

Always.

The Limit of a Function

Concepts to Master

A. Limits; Numerical estimates; Graphical estimates
B. Right- and left-hand limits; Relationships between limits and one-sided limits

Summary and Focus Questions

A. Intuitively, $\lim_{x \to a} f(x) = L$ means that as x gets closer and closer to the number a (but not equal to a), the corresponding $f(x)$ values get closer and closer to the number L. For now, ignore the value $f(a)$, if there is one, in determining a limit.

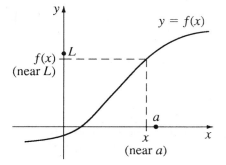

To evaluate $\lim_{x \to a} f(x)$ from the graph of $y = f(x)$, simply note what corresponding $f(x)$ values approach on the y-axis as x values are chosen near a on the x-axis.

A somewhat risky method of calculating $\lim_{x \to a} f(x)$ is to select several values of x near a and observe the pattern of corresponding $f(x)$ values. Your answer may vary depending on the particular x values you choose.

Section 1.2: The Limit of a Function

1) Answer each using the graph below:

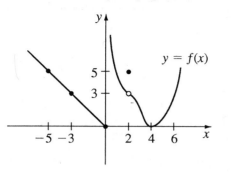

$y = f(x)$

a) $\lim\limits_{x \to -5} f(x) =$

5.

b) $\lim\limits_{x \to 2} f(x) =$

3.

c) $\lim\limits_{x \to 0} f(x) =$

does not exist.

d) $\lim\limits_{x \to 4} f(x) =$

0.

2) Estimate $\lim\limits_{x \to 2} |x - 5|$.

We make a table of values:

x	$f(x) = |x - 5|$
3	2
2.1	2.9
1.99	3.01
2.0003	2.9997

3) Estimate $\displaystyle\lim_{x \to 1} \frac{1}{1-x}$.

We guess that $\displaystyle\lim_{x \to 2} \left| x - 5 \right| = 3$.

We make a table of values:

x	$f(x) = \dfrac{1}{1-x}$
1.5	2
1.01	-100
.99	100
1.0004	-2500

The values of $f(x)$ do not seem to congregate near a particular value. We guess that

$\displaystyle\lim_{x \to 1} \frac{1}{1-x}$ <u>does not exist.</u>

B. $\displaystyle\lim_{x \to a^-} f(x) = L$, <u>the left-hand limit of f at a</u>, means that $f(x)$ gets closer and closer to L as x approaches a and $x < a$.

$\displaystyle\lim_{x \to a^+} f(x) = L$, <u>the right-hand limit</u>, is the same concept but involves only values of x greater than a.

If $\displaystyle\lim_{x \to a^-} f(x) = L$ and $\displaystyle\lim_{x \to a^+} f(x) = L$ both exist and are the same number L, then $\displaystyle\lim_{x \to a} f(x)$ exists and is L. If either one-sided limit doesn't exist or if they both exist but are different numbers, then $\displaystyle\lim_{x \to a} f(x)$ does not exist. Conversely if $\displaystyle\lim_{x \to a} f(x) = L$, then both one-sided limits equal L.

4) Answer each using the graph below:

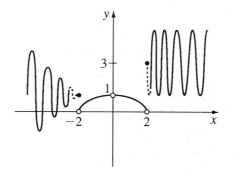

a) $\lim\limits_{x \to -2^-} f(x) =$ _____.

b) $\lim\limits_{x \to 2^+} f(x) =$ _____.

c) $\lim\limits_{x \to 2^-} f(x) =$ _____.

d) $\lim\limits_{x \to 2} f(x) =$ _____.

e) $\lim\limits_{x \to 0} f(x) =$ _____.

5) $\lim\limits_{x \to 3^-} \dfrac{3 - x}{|x - 3|} =$ _____.

1.

does not exist.

0.

does not exist.

1.

1. For x near 3 with $x < 3$, $x - 3 < 0$.

Thus $|x - 3| = -(x - 3) = 3 - x$.

Therefore $\lim\limits_{x \to 3^-} \dfrac{3 - x}{|x - 3|} =$

$\lim\limits_{x \to 3^-} \dfrac{3 - x}{3 - x} = 1.$

6) Find $\lim\limits_{x \to 2} f(x)$, where

$$f(x) = \begin{cases} 3x + 1 & \text{if } x < 2 \\ 5 & \text{if } x = 2 \\ x^2 + 3 & \text{if } x > 2 \end{cases}$$

$$\lim_{x \to 2^+} f(x) = \lim_{x \to 2^+} x^2 + 3 = 7.$$

$$\lim_{x \to 2^-} f(x) = \lim_{x \to 2^-} 3x + 1 = 7.$$

Since both one-sided limits are 7,

$$\lim_{x \to 2} f(x) = 7.$$

7) Sometimes, Always, or Never :

If $\lim\limits_{x \to 2^-} f(x)$ does not exist, then $\lim\limits_{x \to 2^+} f(x)$ does not exist.

Sometimes.

True in the case $f(x) = \dfrac{1}{x - 2}$.

False in the case $f(x) = \dfrac{x - 2}{|x - 2|}$.

8) Sometimes, Always, or Never :

If $\lim\limits_{x \to 2^+} f(x)$ does not exist, then $\lim\limits_{x \to 2} f(x)$ does not exist.

Always.

Calculating Limits Using the Limit Laws

Concepts to Master

A. Evaluation of limits of rational functions at points in their domain
B. Limit laws (theorems) for evaluating limits
C. Algebraic manipulations prior to using the limit laws
D. Squeeze theorem

Summary and Focus Questions

A. Limits of *some* functions at *some* points are easy to find. If $f(x)$ is rational and a is in the domain of f, then $\lim_{x \to a} f(x) = f(a)$.

For example, $\lim_{x \to 3} \dfrac{x}{x+2} = \dfrac{3}{3+2} = \dfrac{3}{5}$ since 3 is in the domain of $f(x)$.

Since polynomials are rational functions, the above applies to them as well, as in
$\lim_{x \to 2} (4x^3 - 6x^2 + 10x + 2) = 4(2)^3 - 6(2)^2 + 10(2) + 2 = 30.$

1) Find $\lim_{x \to 6} (4x^2 - 10x + 3)$.

$4(6)^2 - 10(6) + 3 = 87$.

2) Find $\lim_{x \to 1} \dfrac{x^2 - 6x + 5}{x - 1}$.

Since 1 is not in the domain, we cannot substitute $x = 1$ to find this limit. The substitution yields $\dfrac{0}{0}$ which is a signal that the limit might still exist but must be found by a different method (see question 8.)

B. The eleven basic limit laws of this section, when applicable, allow you to evaluate complex limits by rewriting them as algebraic combinations of simpler limits of rational functions, as in:

$$\lim_{x\to 3}\sqrt{x^2+x} = \sqrt{\lim_{x\to 3}(x^2+x)} = \sqrt{\lim_{x\to 3}x^2 + \lim_{x\to 3}x} = \sqrt{9+3} = \sqrt{12}.$$

The above was carried out to extremes to demonstrate the limit laws; from part **A** we could have concluded directly that $\lim_{x\to 3}(x^2+x) = 12$.

Most of the time the limit laws fail when a zero denominator results (see part **C**).

3) True or False:

$$\lim_{x\to a}(f(x)g(x)) = \lim_{x\to a}f(x)\lim_{x\to a}g(x).$$

<u>True</u>, provided the limits exist.

4) True or False:

$$\lim_{x\to a}f(g(x)) = \lim_{x\to a}f(x)\lim_{x\to a}g(x).$$

<u>False</u>.

5) Evaluate $\lim\limits_{x\to 5}\sqrt[3]{\dfrac{4x+44}{6x-29}}$.

$$\lim_{x\to 5}\sqrt[3]{\frac{4x+44}{6x-29}} = \sqrt[3]{\lim_{x\to 5}\frac{4x+44}{6x-29}}$$

$$= \sqrt[3]{\frac{64}{1}} = 4.$$

6) Evaluate $\lim\limits_{x\to 1}\dfrac{x^2(6x+3)(2x-7)}{(x^3+4)(x+17)}$.

Rather than multiplying this function out, this limit is

$$\left| \begin{array}{l} \dfrac{\left(\lim\limits_{x\to1} x^2\right)\left(\lim\limits_{x\to1} (6x+3)\right)\left(\lim\limits_{x\to1} (2x-7)\right)}{\left(\lim\limits_{x\to1} (x^3+4)\right)\ \left(\lim\limits_{x\to1}(x+17)\right)} \\[20pt] = \dfrac{(1)(9)(-5)}{(5)(18)} = -\dfrac{1}{2} \ . \end{array} \right.$$

C. There are many times that direct substitution and the limit laws fail, as in:

$$\lim_{x\to4} \frac{x^2-16}{x-4} \ \text{(gives } \frac{0}{0}, \text{ no information).}$$

To evaluate a limit first try the limit laws. If they fail, you should next resort to algebraic techniques to rewrite the function in a form for which the limit theorems do work.

Division by zero is a common reason why the limit laws fail. You should try to rewrite the function in a form that will not give a zero denominator. There are many techniques for this and often more than one technique is required in a given problem. The most common are:

Cancellation:

Frequently when substitution of $x = a$ in $\lim\limits_{x\to a} \dfrac{f(x)}{g(x)}$ results in $\frac{0}{0}$, $x - a$ will be a factor of both $f(x)$ and $g(x)$ and hence can be canceled.

Find $\lim\limits_{x\to2} \dfrac{2x^2-4x}{x^2-4}$.

Here $x = 2$ leaves $\frac{0}{0}$ but $\dfrac{2x^2-4x}{x^2-4} = \dfrac{2x(x-2)}{(x+2)(x-2)} = \dfrac{2x}{x+2}$, provided $x \neq 2$.

Thus $\lim\limits_{x\to2} \dfrac{2x^2-4x}{x^2-4} = \lim\limits_{x\to2} \dfrac{2x}{x+2} = \dfrac{2\cdot2}{2+2} = 1.$

Fraction Manipulation:

Some limits involve complex fractions, which when simplified can be solved using limit laws.

Find $\lim\limits_{x \to 1} \dfrac{\frac{1}{x} - x}{\frac{1}{x} - 1}$.

We rewrite the function by first finding a simpler numerator and denominator and then using cancellation:

$$\frac{\frac{1}{x} - x}{\frac{1}{x} - 1} = \frac{\frac{1}{x} - \frac{x^2}{x}}{\frac{1}{x} - 1} = \frac{\frac{1-x^2}{x}}{\frac{1-x}{x}} = \frac{1 - x^2}{1 - x} = \frac{(1 - x)(1 + x)}{1 - x} = 1 + x, \text{ for } x \neq 1.$$

Thus $\lim\limits_{x \to 1} \dfrac{\frac{1}{x} - x}{\frac{1}{x} - 1} = \lim\limits_{x \to 1} (1 + x) = 2.$

Rationalizing an Expression:

Frequently if one or more radicals appear in the function, a process similar to using the conjugate of a complex number may be used to rewrite part of the function without radicals. For functions with $\sqrt{a} - \sqrt{b}$, use $\dfrac{\sqrt{a} + \sqrt{b}}{\sqrt{a} + \sqrt{b}}$. The result may then be simplified using cancellation.

Find $\lim\limits_{x \to 0} \dfrac{\sqrt{2 + x} - \sqrt{2}}{x}$.

Here $\frac{0}{0}$ results, but if $\sqrt{2 + x} - \sqrt{2}$ is multiplied by $\sqrt{2 + x} + \sqrt{2}$, no radicals will remain in the numerator:

$$\frac{\sqrt{2 + x} - \sqrt{2}}{x} \cdot \frac{\sqrt{2 + x} + \sqrt{2}}{\sqrt{2 + x} + \sqrt{2}} = \frac{2 + x - 2}{x(\sqrt{2 + x} + \sqrt{2})} = \frac{1}{\sqrt{2 + x} + \sqrt{2}}.$$

$$\lim_{x \to 0} \frac{\sqrt{2+x} - \sqrt{2}}{x} = \lim_{x \to 0} \frac{1}{\sqrt{2+x} + \sqrt{2}} = \frac{1}{\sqrt{2} + \sqrt{2}} = \frac{1}{2\sqrt{2}}.$$

7) Evaluate $\lim_{x \to 6} \dfrac{2x - 12}{x^2 - x - 30}$.

$$\frac{2x - 12}{x^2 - x - 30} = \frac{2(x - 6)}{(x - 6)(x + 5)}$$

$$= \frac{2}{x + 5}, \text{ for } x \neq 6 .$$

$$\lim_{x \to 6} \frac{2}{x + 5} = \frac{2}{11}.$$

8) Evaluate $\lim_{x \to 1} \dfrac{x^2 - 6x + 5}{x - 1}$.

$$\frac{x^2 - 6x + 5}{x - 1} = \frac{(x - 1)(x - 5)}{x - 1}$$

$$= x - 5, \text{ for } x \neq 1.$$

$$\lim_{x \to 1} x - 5 = -4.$$

9) Evaluate $\lim_{x \to 3} \left[\dfrac{2x^2}{x - 3} + \dfrac{6x}{3 - x} \right]$.

$$\frac{2x^2}{x - 3} + \frac{6x}{3 - x} = \frac{2x^2}{x - 3} - \frac{6x}{x - 3}$$

$$= \frac{2x^2 - 6x}{x - 3} = \frac{2x(x - 3)}{x - 3} = 2x, \text{ for } x \neq 3.$$

$$\lim_{x \to 3} 2x = 6.$$

Section 1.3: Calculating Limits Using the Limit Laws

10) Evaluate $\lim\limits_{x \to 0} \dfrac{\frac{1}{4+x} - \frac{1}{4}}{x}$.

$$\frac{\frac{1}{4+x} - \frac{1}{4}}{x} = \frac{\frac{4}{(4+x)4} - \frac{(4+x)}{(4+x)4}}{x}$$

$$= \frac{\frac{4 - (4+x)}{(4+x)4}}{x} = \frac{\frac{-x}{(4+x)4}}{x} = \frac{-1}{(4+x)4}.$$

$$\lim\limits_{x \to 0} \frac{-1}{(4+x)4} = -\frac{1}{16}.$$

11) Evaluate $\lim\limits_{x \to 3} \dfrac{\sqrt{x} - \sqrt{3}}{x - 3}$.

$$\frac{\sqrt{x} - \sqrt{3}}{x - 3} \cdot \frac{\sqrt{x} + \sqrt{3}}{\sqrt{x} + \sqrt{3}}$$

$$= \frac{x - 3}{(x-3)\left(\sqrt{x} + \sqrt{3}\right)}$$

$$= \frac{1}{\sqrt{x} + \sqrt{3}}, \text{ for } x \neq 3.$$

$$\lim\limits_{x \to 3} \frac{1}{\sqrt{x} + \sqrt{3}} = \frac{1}{2\sqrt{3}}.$$

D. If $g(x)$ is "trapped" between $f(x)$ and $h(x)$ (i.e. if $f(x) \leq g(x) \leq h(x)$ for all x near a, except $x = a$) then $\lim\limits_{x \to a} f(x) \leq \lim\limits_{x \to a} g(x) \leq \lim\limits_{x \to a} h(x)$, if they each exist. If it happens that $\lim\limits_{x \to a} f(x) = \lim\limits_{x \to a} h(x)$ then $\lim\limits_{x \to a} g(x)$ must exist and be that same number (the Squeeze Theorem).

12) True or False:

If $f(x) < g(x)$ for all x then
$$\lim_{x \to a} f(x) < \lim_{x \to a} g(x).$$

False. Let $a = 0$, $f(x) = -x^2$

and $g(x) = \begin{cases} x^2 & \text{if } x \neq 0 \\ 1 & \text{if } x = 0 \end{cases}$.

$f(x) < g(x)$, for all x but
$$\lim_{x \to 0} -x^2 = 0 = \lim_{x \to 0} g(x).$$

13) For $0 \le x \le 1$, $x + 1 \le 3^x \le 2x + 1$.

Use this to find $\lim_{x \to 0^+} 3^x$.

Since $\lim_{x \to 0^+} (x + 1) = 1$

and $\lim_{x \to 0^+} (2x + 1) = 1$

by the Squeeze Theorem, $\lim_{x \to 0^+} 3^x = 1$.

The Precise Definition of a Limit

Concepts to Master

Epsilon-delta (ε-δ) definition of $\lim\limits_{x \to a} f(x) = L$

Summary and Focus Questions

$\lim\limits_{x \to a} f(x) = L$ means for any given $\varepsilon > 0$, there exists $\delta > 0$ such that

$$\text{if } 0 < |x - a| < \delta, \text{ then } |f(x) - L| < \varepsilon.$$

This is consistent with the intuitive definition given earlier if you think of ε and δ as small numbers and note that $|f(x) - L|$ is the distance between $f(x)$ and L while $|x - a|$ is the distance between x and a. This definition says that $f(x)$ will always be within ε units of L ($|f(x) - L| < \varepsilon$) whenever x is within δ units of a ($|x - a| < \delta$).

A proof that $\lim\limits_{x \to a} f(x) = L$ consists of a verification of the definition. Start by assuming you have an $\varepsilon > 0$ and then determine a number δ so that the statement $|f(x) - L| < \varepsilon$ can be deduced from the statement $0 < |x - a| < \delta$. There are no universal techniques for determining δ in terms of ε but often this is accomplished by starting with $|f(x) - L| < \varepsilon$ and replacing it with equivalent statements until finally the statement $|x - a| < \delta$ results. The reversal of these steps constitutes a major portion of the proof.

Reducing $|f(x) - L| < \varepsilon$ down to $|x - a| < \delta$ generally requires several rewritings of $|f(x) - L| < \varepsilon$ using factoring and cancellation. You also use properties of absolute value, the most common being $|a \cdot b| = |a||b|$, and $|a| < b$ is equivalent to $-b < a < b$.

<u>Example:</u>

To prove $\lim_{x \to 3} (6x - 5) = 13$ we start with $|f(x) - L|$ and simplify:

$$|f(x) - L| = |6x - 5 - 13| = |6x - 18| = |6(x - 3)| = |6||x - 3|$$

$$= 6|x - 3|.$$

Thus $|6x - 5 - 13| < \varepsilon$ is equivalent to $6|x - 3| < \varepsilon$.

Divide by 6: $|x - 3| < \dfrac{\varepsilon}{6}$. Our choice for δ is $\dfrac{\varepsilon}{6}$.

When $f(x)$ is nonlinear, a simplified $|f(x) - L| < \varepsilon$ may be in the form $|M(x)||x - a| < \varepsilon$ where $M(x)$ is some expression involving x. The procedure here is to find a constant K such that $|M(x)| < K$ for all x such that $|x - a| < p$ for some positive number p (often $p = 1$). The choice of δ is then the smaller of p and $\dfrac{\varepsilon}{K}$, written $\delta = \min\{p, \dfrac{\varepsilon}{K}\}$.

<u>Example:</u>

To prove $\lim_{x \to 4} x^2 - 1 = 15$, we have

$$|f(x) - L| = |x^2 - 1 - 15| = |x^2 - 16| = |(x + 4)(x - 4)| = |x + 4||x - 4|.$$

For $p = 1$, if $|x - 4| < p$,

then $|x - 4| < 1$

$$-1 < x - 4 < 1$$
$$-3 < \quad x \quad < 5$$
$$1 < \quad x + 4 \quad < 9$$

Thus $|x + 4| < 9$.

Therefore, using $K = 9$, when $|x - 4| < 1$, we have

$$|f(x) - L| = |x + 4||x - 4| < 9|x - 4|$$

and

$$9|x - 4| < \varepsilon \text{ if } |x - 4| < \frac{\varepsilon}{9}.$$

Thus choose $\delta = \min\{1, \frac{\varepsilon}{9}\}$.

1) What part of the definition of $\lim_{x \to a} f(x) = L$ ensures that $x = a$ is not considered?

$$0 < |x - a|.$$

2) The choice of δ usually (does, does not) depend upon the size of ε.

<u>Does</u>. For most limits, the smaller the ε given, the smaller must be the corresponding choice of δ.

3) For arbitrary $\varepsilon > 0$, find a $\delta > 0$, in terms of ε such that the definition of limit is satisfied for $\lim_{x \to 4} (2x + 1) = 9$.

$|f(x) - L| < \varepsilon$ (Start)

$|2x + 1 - 9| < \varepsilon$ (Replacement)

$|2x - 8| < \varepsilon$ (Simplification)

$2|x - 4| < \varepsilon$ (Factoring)

$|x - 4| < \dfrac{\varepsilon}{2}$ (Divide by 2)

The desired δ is $\delta = \dfrac{\varepsilon}{2}$.

4) Find δ for an arbitrary ε in $\lim\limits_{x \to 4} x^2 - 4 = 5$.

$|f(x) - L| = |x^2 - 4 - 5| = |x^2 - 9|$

$= |(x + 3)(x - 3)| = |x + 3||x - 3|$

If $p = 1$ and $|x - 3| < 1$, then

$-1 < x - 3 < 1$

$2 < x < 4$

$5 < x + 3 < 7$

So $|x + 3| < 7$.

Let $\delta = \min\left\{ 1, \dfrac{\varepsilon}{7} \right\}$.

Continuity

Concepts to Master

A. Continuity of a function at a point and on an interval
B. Intermediate Value Theorem

Summary and Focus Questions

A. A function f is <u>continuous at a</u> means both numbers $f(a)$ and $\lim_{x \to a} f(x)$ exist and are equal, that is, $\lim_{x \to a} f(x) = f(a)$.
This makes the evaluation of limits of continuous functions easy —just substitute the value $x = a$ in $f(x)$.

f is <u>continuous on an interval</u> means f is continuous at each point in the interval. The graph of $f(x)$ must be unbroken at each point in the interval.

Algebraic (and therefore rational and polynomial) functions are continuous at each point in their domains.

1)

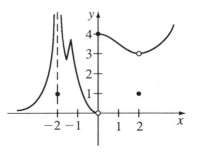

Determine the points where the function f
above is not continuous (*is* discontinuous).

Discontinous at

$x = -2$, because $\lim\limits_{x \to -2} f(x)$ does not exist.

$x = 0$, because $\lim\limits_{x \to 0} f(x)$ does not exist.

$x = 2$, because $f(2) = 1$, but $\lim\limits_{x \to 2} f(x) = 3$.

2) For the graph in problem 1 determine
   whether f is continuous on:

   a) $[-4, -3]$

   Yes.

   b) $[5, 1.5]$

   Yes.

   c) $[-1, 1]$

   No, not continuous at 0.

   d) $[-2, -1]$

   Yes.

   e) $[0, 1]$

   Yes.

3) Where is $f(x) = \dfrac{x}{(x-1)(x+2)}$ continuous?

   Since f is rational, f is continuous
   everywhere it is defined:
   continous for all x except $x = 1$, $x = -2$.

4) Is $f(x) = 4x^2 + 10x + 1$ continuous
   at $x = \sqrt{2}$?

   Yes. It is a polynomial, so f is continuous
   everywhere.

5) Where is $f(x) = \sqrt{\dfrac{1-x}{x}}$ continuous?

$(-\infty, 0) \cup (0, 1]$, f is algebraic and this is the domain of f.

B. The Intermediate Value Theorem is:

If f is continuous on $[a, b]$ and N is between $f(a)$ and $f(b)$ then there is at least one c between a and b so that $f(c) = N$.

Thus as x varies from a to b, $f(x)$ must at least pick up all values between $f(a)$ and $f(b)$.

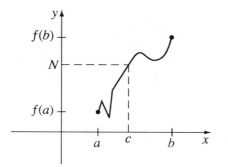

6) Does $f(x) = x^2 - x - 3$ have a zero (a value w such that $f(w) = 0$) in $[1, 3]$?

<u>Yes.</u> Since f is a polynomial f is continuous on $[1, 3]$. Since 0 is between $-3 = f(1)$ and $3 = f(3)$, by the Intermediate Value Theorem, there exists w between 1 and 3 with $f(w) = 0$.

7) Is the hypothesis of the Intermediate Value Theorem satisfied for $f(x) = \begin{cases} \sin \frac{\pi}{x} & x \neq 0 \\ 0 & x = 0 \end{cases}$

with domain $[-2, 2]$?

<u>No.</u> $f(x)$ is not continuous at 0, hence not continuous on $[-2, 2]$.

Limits at Infinity; Horizontal Asymptotes

Concepts to Master

A. Define and evaluate limits at infinity
B. Horizontal asymptotes

Summary and Focus Questions

A. $\lim\limits_{x \to \infty} f(x) = L$ means for each $\varepsilon > 0$ there exists a number N such that $|f(x) - L| < \varepsilon$ whenever $x > N$. Intuitively this means that as x is assigned larger values without bound, the corresponding values of $f(x)$ approach L.

$\lim\limits_{x \to -\infty} f(x) = L$ is defined similarly and means that $f(x)$ approaches L as x takes more negative values without bound.

Most of the techniques for evaluating ordinary limits may be used for limits at infinity. In addition, for a limit involving a rational function, dividing the numerator and denominator by the highest power of x that appears may help, as in:

$$\lim_{x \to \infty} \frac{4x^2 + 7x + 5}{2x^2 + 3} = \lim_{x \to \infty} \frac{\dfrac{(4x^2 + 7x + 5)}{x^2}}{\dfrac{(2x^2 + 3)}{x^2}} = \lim_{x \to \infty} \frac{4 + \dfrac{7}{x} + \dfrac{5}{x^2}}{2 + \dfrac{3}{x^2}} = \frac{4 + 0 + 0}{2 + 0} = 2.$$

All of the usual limit theorems hold for limits at infinity.

1) $\lim\limits_{x\to\infty} f(x) = L$ means for all _____

there exists _____ such

that _____

whenever _____.

$\varepsilon > 0$

N

$|\,f(x) - L\,| < \varepsilon$

$x > N$.

2) Evaluate

a) $\lim\limits_{x\to\infty} \dfrac{5x^3 + 7x + 1}{3x^3 + 2x^2 + 3}$

Divide both numerator and denominator
by x^3.

$$\lim\limits_{x\to\infty} \frac{5 + \dfrac{7}{x^2} + \dfrac{1}{x^3}}{3 + \dfrac{2}{x} + \dfrac{3}{x^3}} = \frac{5 + 0 + 0}{3 + 0 + 0} = \frac{5}{3}.$$

Limits of this type can also be "eye-balled"
by observing that for very large positive x,
$5x^3 + 7x + 1 \approx 5x^3$ and
$3x^3 + 2x^2 + 3 \approx 3x^3$. Thus the limit is

$$\lim\limits_{x\to\infty} \frac{5x^3}{3x^3} = \frac{5}{3}.$$

b) $\lim\limits_{x\to -\infty} \dfrac{2x + 5}{\sqrt{x^2 + 4}}$

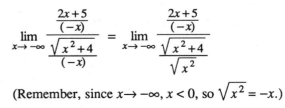

$$\lim\limits_{x\to -\infty} \frac{\dfrac{2x+5}{(-x)}}{\dfrac{\sqrt{x^2+4}}{(-x)}} = \lim\limits_{x\to -\infty} \frac{\dfrac{2x+5}{(-x)}}{\dfrac{\sqrt{x^2+4}}{\sqrt{x^2}}}$$

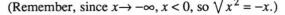

(Remember, since $x\to -\infty$, $x < 0$, so $\sqrt{x^2} = -x$.)

$$= \lim_{x \to -\infty} \frac{-2 - \frac{5}{x}}{\sqrt{1 + \frac{4}{x^2}}} = \frac{-2 - 0}{\sqrt{1 + 0}} = -2.$$

c) $\displaystyle\lim_{x \to \infty} \frac{\cos x}{x}$

Since $-1 \le \cos x \le 1$, $\dfrac{-1}{x} \le \dfrac{\cos x}{x} \le \dfrac{1}{x}$.

Both $\displaystyle\lim_{x \to \infty} \frac{-1}{x} = 0$ and $\displaystyle\lim_{x \to \infty} \frac{1}{x} = 0$.

Therefore $\displaystyle\lim_{x \to \infty} \frac{\cos x}{x} = 0$.

3) True or False:

If all three limits exist, then
$$\lim_{x \to \infty} (f(x)g(x)) = \left(\lim_{x \to \infty} f(x)\right)\left(\lim_{x \to \infty} g(x)\right).$$

True.

B. A <u>horizontal asymptote</u> for a function f is a line $y = L$ such that $\displaystyle\lim_{x \to \infty} f(x) = L$, $\displaystyle\lim_{x \to -\infty} f(x) = L$, or both. A function can have at most two horizontal asymptotes.

Here are the graphs of three examples:

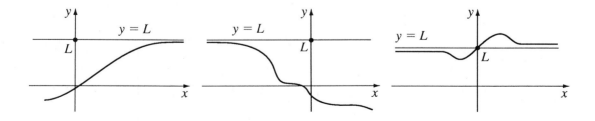

4) Find the horizontal asymptotes for the graph:

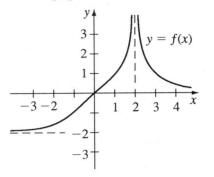

f has two horizontal asymptotes, $y = -2$ and $y = 0$.

5) Find the horizontal asymptotes for:

a) $f(x) = \dfrac{x}{|x| - 1}$

$y = 1$, because $\displaystyle\lim_{x \to \infty} \dfrac{x}{|x| - 1} = 1$

and $y = -1$, because $\displaystyle\lim_{x \to -\infty} \dfrac{x}{|x| - 1} = -1$.

b) $f(x) = \dfrac{x}{x^2 - 1}$

$\displaystyle\lim_{x \to \infty} \dfrac{x}{x^2 - 1} = 0$ and $\displaystyle\lim_{x \to -\infty} \dfrac{x}{x^2 - 1} = 0$,

so $y = 0$ is the only horizontal asymptote.

Infinite Limits; Vertical Asymptotes

Concepts to Master

A. Define and evaluate infinite limits
B. Vertical asymptotes

Summary and Focus Questions

A. $\lim\limits_{x \to a} f(x) = \infty$ means that for all positive M there is $\delta > 0$, such that $f(x) \geq M$ whenever $0 < |x - a| < \delta$. In other words, as x is assigned values that approach a, the corresponding $f(x)$ values grow larger without bound.

$\lim\limits_{x \to a} f(x) = -\infty$ means $f(x)$ becomes smaller without bound as x approaches a.
As with ordinary limits, for $\lim\limits_{x \to a^+} f(x) = \pm\infty$, consider only $x > a$, and for $\lim\limits_{x \to a^-} f(x) = \pm\infty$, consider only $x < a$.

$\lim\limits_{x \to \infty} f(x) = \infty$ means as x increases without bound then so do the corresponding $f(x)$ values.

These limits of ∞ and $-\infty$ are not numbers, so the standard limit theorems do not hold. Factoring, canceling, and other algebraic techniques sometimes help, as in:

$$\lim_{x \to 2^+} \frac{x + 2}{x^2 - 4} = \lim_{x \to 2^+} \frac{x + 2}{(x - 2)(x + 2)} = \lim_{x \to 2^+} \frac{1}{x - 2} = \infty.$$

1) Evaluate the limits given the graph of
   $y = f(x)$ below:

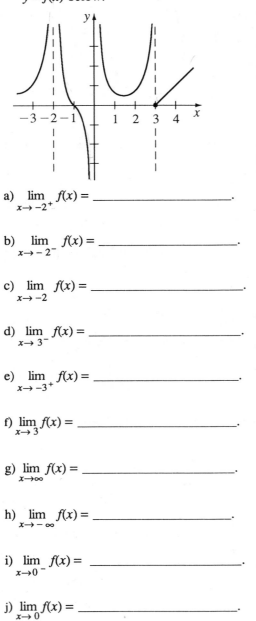

a) $\lim\limits_{x \to -2^+} f(x) =$ _____.

        ∞.

b) $\lim\limits_{x \to -2^-} f(x) =$ _____.

        ∞.

c) $\lim\limits_{x \to -2} f(x) =$ _____.

        ∞.

d) $\lim\limits_{x \to 3^-} f(x) =$ _____.

        ∞.

e) $\lim\limits_{x \to -3^+} f(x) =$ _____.

        0.

f) $\lim\limits_{x \to 3} f(x) =$ _____.

        does not exist.

g) $\lim\limits_{x \to \infty} f(x) =$ _____.

        ∞.

h) $\lim\limits_{x \to -\infty} f(x) =$ _____.

        0.

i) $\lim\limits_{x \to 0^-} f(x) =$ _____.

        $-\infty$.

j) $\lim\limits_{x \to 0} f(x) =$ _____.

        does not exist.

2) $\displaystyle\lim_{x\to 3}\frac{1}{|x-3|} = $ _____.

∞. For x near 3, but not equal to 3, $|x-3|$ is a small positive number.

Therefore, $\dfrac{1}{|x-3|}$ is a large positive number.

3) True or False:

If $\displaystyle\lim_{x\to a} f(x) = \infty$, and $\displaystyle\lim_{x\to a} g(x) = \infty$, then $\displaystyle\lim_{x\to a}(f(x) - g(x)) = 0$.

<u>False</u>. For example, $f(x) = \dfrac{3}{x^2}$ and

$g(x) = \dfrac{2}{x^2}$. Then $\displaystyle\lim_{x\to 0}\frac{3}{x^2} = \infty$,

$\displaystyle\lim_{x\to 0}\frac{2}{x^2} = \infty$, but $\displaystyle\lim_{x\to 0}\left(\frac{3}{x^2} - \frac{2}{x^2}\right) =$

$\displaystyle\lim_{x\to 0}\frac{1}{x^2} = \infty$ (not 0 !).

4) $\displaystyle\lim_{x\to 3}\frac{x}{3-x}$

For x near 3, $\dfrac{x}{3-x}$ is the result of a number

near 3 divided by a number near 0. Consider one-sided limits:

$\displaystyle\lim_{x\to 3^+}\frac{x}{3-x} = -\infty$ because the denominator

is negative, while $\displaystyle\lim_{x\to 3^-}\frac{x}{3-x} = \infty$.

Thus $\lim\limits_{x\to 3}\dfrac{x}{3-x}$ does not exist.

5) $\lim\limits_{x\to\infty}\dfrac{6-2x^2}{x+7}$

Dividing by the highest power of x in the

denominator gives $\lim\limits_{x\to\infty}\dfrac{\dfrac{6-2x^2}{x}}{\dfrac{x+7}{x}} =$

$\lim\limits_{x\to\infty}\dfrac{\dfrac{6}{x}-2x}{1+\dfrac{7}{x}}$. As $x\to\infty$, $1+\dfrac{7}{x}\to 1$

while $\dfrac{6}{x}-2x$ becomes a large negative

number. Thus $\lim\limits_{x\to\infty}\dfrac{6-2x^2}{x+7}=-\infty$.

B. A <u>vertical asymptote</u> for $y=f(x)$ is a vertical line $x=a$ for which at least one of these limits hold:

$$\lim_{x\to a^-} f(x) = \infty, \qquad \lim_{x\to a^-} f(x) = -\infty, \qquad \lim_{x\to a^+} f(x) = \infty, \qquad \lim_{x\to a^+} f(x) = -\infty.$$

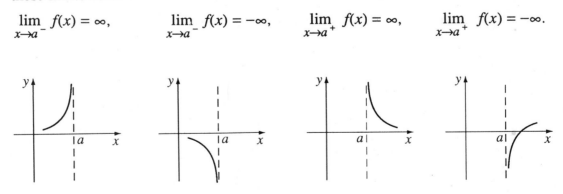

A function may have any number of vertical asymptotes. Some, like $y=\tan x$ have an infinite number.

6) What are the vertical asymptotes for the function in question 1?

$x = -2$, $x = 0$, and $x = 3$.

7) Find the vertical asymptotes of

$$f(x) = \frac{x}{|x| - 1}.$$

Good places to check are where $f(x)$ is undefined: at 1 and -1.

$$\lim_{x \to 1^+} \frac{x}{|x| - 1} = \infty, \text{ and}$$

$$\lim_{x \to -1^-} \frac{x}{|x| - 1} = \infty,$$

so $x = 1$ and $x = -1$ are asymptotes.

8) Find the vertical asymptotes of

$$f(x) = \frac{x - x^2}{x}.$$

$f(x)$ is not defined for $x = 0$, so $x = 0$ could be a vertical asymptote. However,

$$\lim_{x \to 0} \frac{x - x^2}{x} = \lim_{x \to 0} 1 - x = 1$$

so f has no vertical asymptotes.

Tangents, Velocities, and Other Rates of Change

Concepts to Master

A. Slope of the tangent line to a graph
B. Instantaneous velocity; Rates of change

Summary and Focus Questions

A. If $m = \lim\limits_{x \to a} \dfrac{f(x) - f(a)}{x - a}$ exists, we say m is the <u>slope of the tangent line</u> to $y = f(x)$

at the point $(a, f(a))$.

If we use $a + h = x$, then as $x \to a$, we have $h \to 0$. Thus an alternate way to write the above limit is

$$\lim_{x \to a} \frac{f(x) - f(a)}{x - a} = \lim_{h \to 0} \frac{f(a + h) - f(a)}{h}.$$

Either may be used to calculate the slope of the tangent line. Notice that the direct substitution $x = a$ or $h = 0$ in these limits always yields $\frac{0}{0}$, so you can never simply use limit laws to evaluate the slope of the tangent line.

1) Find the slope of the tangent line to
$f(x) = x^2 + 2x$ at $x = 3$.

At $x = 3, f(3) = 3^2 + 2(3) = 15$.

$\dfrac{f(x) - f(a)}{x - a} = \dfrac{x^2 + 2x - 15}{x - 3}$

$= \dfrac{(x - 3)(x + 5)}{x - 3}$

$= x + 5$ for $x \neq 3$.

Thus $\lim\limits_{x \to 3} \dfrac{f(x) - f(3)}{x - 3} = \lim\limits_{x \to 3} (x - 5) = 8$.

2) Find $\lim\limits_{h \to 0} \dfrac{f(a + h) - f(a)}{h}$ where

$f(x) = \dfrac{1}{x - 1}$ and $a = 2$.

$f(2 + h) = \dfrac{1}{2 + h - 1} = \dfrac{1}{h + 1}$.

$f(2 + h) - f(2) = \dfrac{1}{h + 1} - \dfrac{1}{2 - 1}$

$= \dfrac{1}{h + 1} - \dfrac{1}{1}$

$= \dfrac{1}{h + 1} - \dfrac{h + 1}{h + 1}$

$= \dfrac{1 - (h + 1)}{h + 1} = \dfrac{-h}{h + 1}$.

Thus $\lim\limits_{h \to 0} \dfrac{f(a + h) - f(a)}{h} = \lim\limits_{h \to 0} \dfrac{-h}{h + 1} \cdot \dfrac{1}{h}$

$= \lim\limits_{h \to 0} \dfrac{-1}{h + 1} = -1$.

B. The expression $\dfrac{f(x) - f(a)}{x - a}$ is the <u>average rate of change</u> of $y = f(x)$ on the

interval between x and a. Thus $\lim\limits_{x \to a} \dfrac{f(x) - f(a)}{x - a}$ is the <u>instantaneous rate of</u>

<u>change</u> of $y = f(x)$ at $x = a$.

In particular, if $y = f(x)$ is the position of an object at time x, $\dfrac{f(x) - f(a)}{x - a}$ is the

<u>average velocity</u> and $\lim\limits_{x \to a} \dfrac{f(x) - f(a)}{x - a}$ is the <u>instantaneous velocity</u> at $x = a$.

3) Find the instantaneous velocity at time
   3 seconds if a particle's position at time
   t is $f(t) = t^2 + 2t$ feet.

<u>8 ft/sec</u>. This question is asking for
the same information as question 1. Here
the rate of change is interpreted as
velocity instead of slope.

2

Derivatives

Derivatives

Concepts to Master

A. Definition of the derivative; Notations
B. Interpretations of the derivative
C. Differentiability; Relationship of continuity to differentiability

Summary and Focus Questions

A. The <u>derivative</u> of $y = f(x)$ at a point x is

$$f'(x) = \lim_{h \to 0} \frac{f(x + h) - f(x)}{h}.$$

All of these notations mean the same thing as $f'(x)$:

$$y', \frac{df}{dx}, \frac{dy}{dx}, \frac{d}{dx} f(x), \ Df(x), \ D_x f(x)$$

For now you must calculate $f'(x)$ in the same manner as the limits of Section 1.8:

(1) Find $f(x + h)$.

(2) Find $f(x + h) - f(x)$.

(3) Determine $\dfrac{f(x + h) - f(x)}{h}$ and simplify (if possible) when $h \neq 0$.

(4) Find the limit of the result of part (3).

1) Find $f'(3)$ for $f(x) = x^2 + 10x$.

$f'(3)$ is found in steps:
(1) $f(3 + h) = (3 + h)^2 + 10(3 + h)$
$= 9 + 6h + h^2 + 30 + 10h$
$= 39 + 16h + h^2$.

2) Find $f'(x)$ for $f(x) = 2x - x^2$.

$(2)\, f(3 + h) - f(3) = (39 + 16h + h^2) - 39$
$= 16h + h^2 = h(16 + h).$

$(3)\, \dfrac{f(3 + h) - f(3)}{h} = \dfrac{h(16 + h)}{h} = 16 + h$

for $h \neq 0$.

(4) Finally, $f'(3) = \lim_{h \to 0}(16 + h) = 16.$

$(1)\ f(x + h) = 2(x + h) - (x + h)^2$
$= 2x + 2h - x^2 - 2xh - h^2.$

$(2)\, f(x + h) - f(x)$
$= 2x + 2h - x^2 - 2xh - h^2 - (2x - x^2)$
$= 2h - 2xh - h^2 = h(2 - 2x - h).$

$(3)\, \dfrac{f(x + h) - f(x)}{h} = \dfrac{h(2 - 2x - h)}{h}$

$= 2 - 2x - h$ for $h \neq 0.$

$(4)\, f'(x) = \lim_{h \to 0}(2 - 2x - h) = 2 - 2x - 0$
$= 2 - 2x.$

3) True or False:

For $y = f(x)$ the notation $f'(x)$

and $\dfrac{dx}{dy}$ have the same meaning.

<u>False.</u> $f'(x) = \dfrac{dy}{dx}$.

B. If $f'(a)$ exists, the tangent line to the graph of f at a is the line through $(a, f(a))$ with slope $f'(a)$. If $f'(a)$ does not exist, then the tangent line might not exist, might be a vertical line, or might not be unique.

The slope of the tangent line (measured by $f'(a)$) is the same as the instantaneous rate of change of $y = f(x)$ with respect to x.

4) The slope of the tangent line to the graph
   of f at the point a is _____.

$f'(a)$, provided it exists.

5) True or False:
   $f'(x)$ measures the average rate of change
   of $y = f(x)$ with respect to x.

False. $f'(x)$ measures <u>instantaneous</u> rate of change.

6) Given that $f'(x) = 3x^2 + 2$
   for $f(x) = x^3 + 2x + 1$, find:

   a) the equation of the tangent line to
   f at the point corresponding to $x = 2$.

The point of tangency has x coordinate 2
and y coordinate $f(2) = 2^3 + 2(2) + 1 = 13$.
The slope is $f'(2) = 3(2)^2 + 2 = 14$.
Hence the tangent line has equation
$y - 13 = 14(x - 2)$ or $y = 14x - 15$.

   b) the instantaneous rate of change of
   $f(x)$ with respect to $x = 3$.

$f'(3) = 3(3)^2 + 2 = 29$.

   c) If $f(x)$ represents the distance in feet
   of a particle from the origin at time x,
   find the velocity at time $x = 4$ sec.

$f'(4) = 3(4)^2 + 3 = 51$ ft /sec.

C. If f is differentiable at a point c, then f is also continuous at c. (Roughly speaking,
   this means that if you can draw a tangent line to a graph, then the graph must be
   unbroken.) The converse is *false:* continuity *does not* imply differentiability.

7) Can a function be continuous at $x = 3$ but not differentiable at $x = 3$?

Yes. $f(x) = |x - 3|$ is an example.

$f'(3)$ does not exist.

8) Can a function be differentiable at $x = 3$ and not continuous at $x = 3$?

No. This can never happen.

9) True or False:

For the function f graphed below:

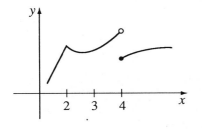

a) $f(x)$ is continuous at 2.

True.

b) $f(x)$ is differentiable at 2.

False.

c) $f(x)$ is continuous at 3.

True.

d) $f(x)$ is differentiable at 3.

True.

e) $f(x)$ is continuous at 4.

False.

f) $f(x)$ is differentiable at 4.

False.

Differentiation Formulas

Concepts to Master

Differentiation from formulas rather than from limits

Summary and Focus Questions

The following differentiation rules <u>must</u> be memorized:

$$\frac{dc}{dx} = 0 \text{ for any constant } c.$$

$$\frac{d}{dx}(x^n) = nx^{n-1} \text{ for any real number. (General Power Rule).}$$

$$[\, c\, f(x)\,]' = c\, f'(x) \text{ for any constant } c.$$

$$[\, f(x) \pm g(x)\,]' = f'(x) \pm g'(x).$$

$$(f(x) \cdot g(x))' = f(x)g'(x) + g(x)f'(x).$$

$$\left[\frac{f(x)}{g(x)}\right]' = \frac{g(x)f'(x) - f(x)g'(x)}{(g(x))^2}.$$

1) Find $f'(x)$ for each:
   a) $f(x) = 10$

   $f'(x) = 0.$

   b) $f(x) = x^6$

   $f'(x) = 6x^5.$

c) $f(x) = 5x^4$

$$f'(x) = 20x^3.$$

d) $f(x) = x^4 + 4x^3$

$$f'(x) = 4x^3 + 12x^2.$$

e) $f(x) = 7x^3 + 6x^2 + 10x + 12$

$$f'(x) = 21x^2 + 12x + 10.$$

f) $f(x) = (x^2 + x)(3x + 1)$

$$f'(x) = (x^2 + x)(3) + (3x + 1)(2x + 1)$$
$$= 9x^2 + 8x + 1.$$

g) $f(x) = \dfrac{x^3}{x^2 + 10}$

$$f'(x) = \frac{(x^2 + 10)(3x^2) - (x^3)(2x)}{(x^2 + 10)^2}$$

$$= \frac{x^4 + 30x^2}{(x^2 + 10)^2}.$$

h) $f(x) = x^{-7}$

$$f'(x) = -7x^{-8}$$

i) $f(x) = x^{\pi}$

$$f'(x) = \pi x^{\pi - 1}$$

j) $f(x) = \dfrac{1 + x^{-1}}{2 - x^{-2}}$

$$f'(x)$$

$$= \frac{(2 - x^{-2})(-x^{-2}) - (1 + x^{-1})(2x^{-3})}{(2 - x^{-2})^2}.$$

Rates of Change in the Natural and Social Sciences

Concepts to Master

Interpretation of a derivative as an instantaneous rate of change of a given quantity

Summary and Focus Questions

For $y = f(x), f'(x)$ measures the instantaneous rate of change of y with respect to x. Rates of change play an important role in many fields. If $f(x)$ represents some quantity at time $x, f(x)$ will measure its instantaneous rate of change at a given time x.

1) Water is flowing out of a water tower in such a fashion that after t minutes there are $10,000 - 10t - t^3$ gallons left. How fast is the water flowing after 2 minutes ?

Let $V(t) = 10,000 - 10t - t^3$ be the volume at time t. The question asks for the rate of change of V:
$V'(t) = -10 - 3t^2$.
At $t = 2$, $V'(2) = -10 - 3(2)^2$
$= -22$ gals/min.

2) A space shuttle is $16t + t^3$ meters from its launch pad t seconds after liftoff. What is its velocity after 3 sec?

$d(t) = 16t + t^3, d'(t) = 16 + 3t^2$
$d'(3) = 16 + 3(3)^2 = 43$ m/sec.

Derivatives of Trigonometric Functions

Concepts to Master

A. Limits involving trigonometric functions
B. Derivatives of the six trigonometric functions

Summary and Focus Questions

A. Two very important limits involving sine and cosine are:

$$\lim_{x \to 0} \frac{\sin x}{x} = 1 \quad \text{and} \quad \lim_{x \to 0} \frac{\cos x - 1}{x} = 0$$

where x is the radian measure of an angle.

Many other limits involving trigonometric functions can be solved when transformed by algebra and limit laws into combinations of these two limits.

Example:

Evaluate $\lim_{t \to 0} \dfrac{\sin 3t}{4t}$.

Since this resembles $\dfrac{\sin x}{x}$, let $x = 3t$. Then as $t \to 0$, $x \to 0$ and

$$\lim_{t \to 0} \frac{\sin 3t}{4t} = \lim_{t \to 0} \left(\frac{3}{4}\right)\frac{\sin 3t}{3t} = \lim_{x \to 0} \left(\frac{3}{4}\right)\frac{\sin x}{x} = \left(\frac{3}{4}\right)1 = \frac{3}{4}.$$

Section 2.4: Derivatives of Trigonometric Functions

1) $\displaystyle\lim_{x\to 0} \frac{1-\cos x}{x} =$ _____.

$\qquad\qquad\qquad$ 0.

2) Find $\displaystyle\lim_{x\to 0} \frac{1}{x\cot x}$.

$$\frac{1}{x\cot x} = \frac{1}{x\dfrac{\cos x}{\sin x}} = \frac{\sin x}{x\cos x}$$

$$= \frac{\sin x}{x} \cdot \frac{1}{\cos x}.$$

Then $\displaystyle\lim_{x\to 0} \frac{\sin x}{x} \cdot \frac{1}{\cos x} = 1\cdot\frac{1}{1} = 1.$

3) Find $\displaystyle\lim_{x\to 0} \frac{x^2}{2\sin x}$.

$$\lim_{x\to 0} \frac{x^2}{2\sin x} = \lim_{x\to 0} \frac{x}{2}\cdot\frac{x}{\sin x}$$

$$= \lim_{x\to 0} \frac{x}{2}\cdot\lim_{x\to 0}\frac{x}{\sin x} = (0)(1) = 0.$$

B. These differentiation formulas <u>must</u> be memorized:

$$\frac{d}{dx}\sin x = \cos x \qquad\qquad \frac{d}{dx}\cos x = -\sin x$$

$$\frac{d}{dx}\tan x = \sec^2 x \qquad\qquad \frac{d}{dx}\cot x = -\csc^2 x$$

$$\frac{d}{dx}\sec x = (\sec x)(\tan x) \qquad\qquad \frac{d}{dx}\csc x = -(\csc x)(\cot x)$$

4) Find y' for each:

a) $y = \sin x - \cos x$

$$y' = \cos x + \sin x.$$

b) $y = \dfrac{\tan x}{x+1}$

$$y' = \frac{(x+1)(\sec^2 x) - (\tan x)(1)}{(x+1)^2}$$

$$= \frac{x \sec^2 x + \sec^2 x - \tan x}{(x+1)^2}.$$

c) $y = \sin \dfrac{\pi}{4}$

$y' = 0$, since y is a constant.

d) $y = x^3 \sin x$

$$y' = x^3(\cos x) + (\sin x)3x^2$$
$$= x^3 \cos x + 3x^2 \sin x.$$

e) $y = x^2 + 2x \cos x$

$$y' = 2x + 2x(-\sin x) + 2 \cos x$$
$$= 2(x - x \sin x + \cos x).$$

The Chain Rule

Concepts to Master

Chain Rule for computing the derivative of a composition

Summary and Focus Questions

If $y = f(u)$ and $u = g(x)$ (and thus $y = f \circ g(x)$) then the <u>Chain Rule</u> for computing the derivative of such a composition is:

$$\frac{dy}{dx} = \frac{dy}{du} \cdot \frac{du}{dx}$$

or equivalently $[f(g(x))]' = f'(g(x)) \cdot g'(x)$.

For example, if $y = (x^2 + 6x + 1)^4$, we recognize $f(x) = x^4$ and $g(x) = x^2 + 6x + 1$ as components such that $y = f((g(x))$.
Thus $y' = f'(g(x)) \cdot g'(x) = 4(x^2 + 6x + 1)^3 \cdot (2x + 6)$.

The key to correct usage of the Chain Rule comes from first recognizing that the function to be differentiated is a composite and determining the components. It is often possible to break down a composite in more than one way. You should use components that are easily differentiated. Composite functions often occur as expressions raised to a power, or as some function (such as log, sin, etc.) of an expression.

<u>Examples:</u>   In each of these, y can be written as $f \circ g(x)$:

$$y = (x + 3)^4 \qquad y = f(u) = u^4 \text{ (the "outer" function)}$$
$$u = g(x) = x + 3 \text{ (the "inner" function)}$$

Section 2.5: The Chain Rule

$$y = \cos x^2 \qquad\qquad y = f(u) = \cos u \,, \, u = g(x) = x^2$$

$$y = \sqrt{1 + \tfrac{1}{x}} \qquad\qquad y = f(u) = \sqrt{u} \,, \, u = g(x) = 1 + \tfrac{1}{x}$$

Sometimes more than two functions are necessary, as in:

$$y = (3 + (x^3 - 2x)^5)^8 \quad y = f(u) = u^8 \text{ (the "outer" function)}$$
$$u = g(v) = 3 + v^5 \text{ (the "middle" function)}$$
$$v = h(x) = x^3 - 2x \text{ (the "inner" function)}.$$

Here $y = f \circ g \circ h(x)$ and

$$\frac{dy}{dx} = 8\underbrace{(3 + (x^3 - 2x)^5)^7}_{\dfrac{dy}{du}} \; \underbrace{(5(x^3 - 2x)^4)}_{\dfrac{du}{dv}} \; \underbrace{(3x^2 - 2)}_{\dfrac{dv}{dx}}$$

1) Find $\dfrac{dy}{dx}$ for:

a) $y = \sqrt{x^3 + 6x}$

Since $y = (x^3 + 6x)^{1/2}$

$y' = \frac{1}{2}(x^3 + 6x)^{-1/2}(3x^2 + 6)$

$= \dfrac{3x^2 + 6}{2\sqrt{x^3 + 6x}}.$

b) $y = \sec x^2$

$y' = (\sec x^2)(\tan x^2)(2x).$

c) $y = \sec^2 x$

Since $y' = [\sec x]^2,$

$$y' = 2[\sec x]^1 (\sec x \tan x)$$
$$= 2 \sec^2 x \, \tan x.$$

d) $y = \cos^3 x^2$

Since $y = [\cos x^2]^3$
$$y' = 3[\cos x^2]^2 \, (-\sin x^2 (2x))$$
$$= -6x \cos^2 x^2 \sin x^2.$$

e) $y = \sin 2x \, \cos 3x$

y is a product of $\sin 2x$ and $\cos 3x$,
so $y' = (\sin 2x)[-\sin 3x \cdot 3]$
$$+ (\cos 3x)[\cos 2x \cdot 2]$$
$$= -3 \sin 2x \sin 3x + 2 \cos 3x \cos 2x.$$

f) $y = \sin(\sec x)$

$$y' = \cos(\sec x) \, (\sec x \, \tan x).$$

g) $y = \tan (x^2 + 1)^4$

$$y' = \sec^2 (x^2 + 1)^4 \, (4(x^2 + 1)^3 (2x))$$
$$= 8x(x^2 + 1)^3 \sec^2 (x^2 + 1)^4.$$

Implicit Differentiation

Concepts to Master

A. Defining a functional relationship implicitly

B. Finding $\dfrac{dy}{dx}$ where the function y of x is given implicitly

Summary and Focus Questions

A. If y is a function of x defined by an equation not of the form $y = f(x)$, we say y as a function of x is defined <u>implicitly</u> by the equation.

<u>Example:</u>

$x + \sqrt{y} = 1$ defines y as a function of x. If we solve for y (and remember that y must be nonnegative) we get $y = (1 - x)^2$ with domain $(-\infty, 1]$.

Equations of relations other than $y = f(x)$ often have graphs that are not the graphs of functions. Subsets of those graphs that *are* graphs of functions will be graphs of functions defined implicitly by the equation.

1) Sketch a graph of $\dfrac{x^2}{9} - \dfrac{y^2}{16} = 1$ and

determine explicit forms for the functions it defines.

We recognize the graph of the equation as a hyperbola opening left and right with asymptotes $y = \pm\dfrac{4}{3}x$.

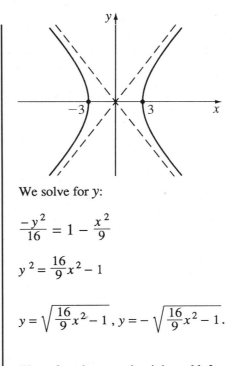

We solve for y:

$$\frac{-y^2}{16} = 1 - \frac{x^2}{9}$$

$$y^2 = \frac{16}{9}x^2 - 1$$

$$y = \sqrt{\frac{16}{9}x^2 - 1} \, , \, y = -\sqrt{\frac{16}{9}x^2 - 1}.$$

These functions are the right and left branches of the hyperbola.

B. Given an equation that defines y as an implicit function of x , $\dfrac{dy}{dx}$ may be found by <u>implicit differentiation</u>:

Take the derivative with respect to x of both sides of the equation, then solve the result for $\dfrac{dy}{dx}$.

Frequently the Chain Rule is used for terms involving y.

2) Find $\dfrac{dy}{dx}$ implicitly:

  a) $x^2 y^3 = 2x + 1$

Differentiate with respect to x (using the

Product Rule because both x^2 and y^3 are functions of x):

$$x^2[3y^2y'] + y^3[2x] = 2.$$

Solve for y':

$$3x^2y^2\,y' = 2 - 2xy^3$$

$$y' = \frac{2 - 2xy^3}{3x^2y^2}.$$

b) $3x^2 - 5xy + y^2 = 10$

Differentiate:

$$6x - (5x(y') + y[5]) + 2y \cdot y' = 0.$$

Solve for y':

$$-5xy' + 2y\,y' = 5y - 6x$$

$$(2y - 5x)y' = 5y - 6x$$

$$y' = \frac{5y - 6x}{2y - 5x}.$$

3) Find the slope of the tangent line to the curve defined by $x^2 + 2xy - y^2 = 1$ at the point (5, 2).

The slope is $\dfrac{dy}{dx}$ at (5, 2). First find $\dfrac{dy}{dx}$

implicitly:

$$2x + (2xy' + y \cdot 2) - 2yy' = 0$$

$$2xy' - 2yy' = -2x - 2y$$

$$2y'(x - y) = -2(x + y)$$

$$y' = -\frac{x + y}{x - y} = \frac{x + y}{y - x}.$$

Now use $x = 5$, $y = 2$:

$$\frac{dy}{dx} = \frac{5 + 2}{2 - 5} = \frac{7}{-3} = -\frac{7}{3}.$$

Higher Derivatives

Concepts to Master

A. Higher derivatives; Higher derivative notations
B. Acceleration as the derivative of velocity

Summary and Focus Questions

A. The <u>second derivative</u> of $y = f(x)$, denoted y'', is the derivative of y'. The <u>third</u> <u>derivative</u>, y''', is the derivative of y''. The nth derivative is denoted $y^{(n)}$.

For $y = x^6$,
$$y' = 6x^5$$
$$y'' = 30x^4$$
$$y''' = 120x^3$$
$$y^{(4)} = 360x^2.$$

All these notations are used for the nth derivative:

$$y^{(n)} \qquad f^{(n)}(x) \qquad \frac{d^n y}{dx^n} \qquad D^n f(x) \qquad D^n y$$

1) Find $y^{(2)}$ for:

a) $y = 5x^3 + 4x^2 + 6x + 3$

$y' = 15x^2 + 8x + 6$, so
$y'' = 30x + 8.$

b) $y = \sqrt{x^2 + 1}$

$y = (x^2 + 1)^{1/2}$

$y' = \frac{1}{2}(x^2 + 1)^{-1/2}(2x)$

$\quad = x \cdot (x^2 + 1)^{-1/2}.$

Thus $y'' =$

$x\left[-\frac{1}{2}(x^2 + 1)^{-3/2}(2x)\right] + (x^2 + 1)^{-1/2}(1)$

$= -x^2(x^2 + 1)^{-3/2} + (x^2 + 1)^{-1/2}.$

2) True or False:

The notation for the sixth derivative of

$y = f(x)$ is $\dfrac{dy^6}{d^6x}$.

<u>False.</u> It is $\dfrac{d^6y}{dx^6}$.

3) Find $f^{(4)}(x)$, where $f(x) = \dfrac{x^{10}}{90} + \dfrac{x^5}{60}$.

$f'(x) = \dfrac{x^9}{9} + \dfrac{x^4}{12}$

$f^{(2)}(x) = x^8 + \dfrac{x^3}{3}$

$f^{(3)}(x) = 8x^7 + x^2$

$f^{(4)}(x) = 56x^6 + 2x$.

4) Find a formula for $f^{(n)}(x)$, where

$f(x) = \dfrac{1}{x^2}.$

$f(x) = x^{-2}$

$f^{(1)}(x) = -2x^{-3}$

$$f^{(2)}(x) = (-2)(-3)x^{-4}$$

$$f^{(3)}(x) = (-2)(-3)(-4)x^{-5}.$$

The pattern is $f^{(n)}(x) =$

$$(-2)(-3)...(-(n+1))x^{-(n+2)}$$

$$= \frac{(-1)^n(n+1)!}{x^{n+2}}.$$

B. The rate of change of velocity is called <u>acceleration.</u> Thus if $y = f(x)$ is a position function, then y'' measures the instantaneous acceleration of the particle.

5) Find the velocity and acceleration after 2 seconds of a particle whose position (in meters) after x seconds is $s = 3t^3 + 6t^2 + t + 1$.

$s' = 9t^2 + 12t + 1$. At $t = 2$, the velocity is $s' = 9(2)^2 + 12(2) + 1 = 61$ m/sec.

$s'' = 18t + 12$. At $t = 2$, the acceleration is $s'' = 18(2) + 12 = 48$ m/sec.

6) Find the acceleration at time $\frac{\pi}{6}$ for a particle whose position in meters at time t seconds is $s = \cos 2t$.

$s = \cos 2t$, $s' = -2 \sin 2t$,

$s'' = -4 \cos 2t$. At $t = \frac{\pi}{6}$,

$s'' = -4 \cos \frac{\pi}{3} = -4\left(\frac{1}{2}\right) = -2$ m/sec.

Related Rates

Concepts to Master

Solutions of related rate problems

Summary and Focus Questions

The procedure to solve a related rate problem is:

(1) Illustrate the problem with a picture if possible.
(2) Identify and label all fixed quantities with constants and all quantities that are functions of time with variables.
(3) Find an equation that relates the constants, the variables whose rates are given, and the variable whose rate is desired.
(4) Differentiate the equation in (3) with respect to time.
(5) Substitute all known quantities in the result of (4) and solve for the unknown rate.

Step (3) is often the hardest and will call upon your skills in remembering geometric and trigonometric relationships.

1) An observer, 300 meters from the launch pad of a rocket, watches it ascend vertically at 60 m/sec. Find the rate of change of the distance between the rocket and the observer when the rocket is 400 meters high.

Section 2.8: Related Rates

(1)

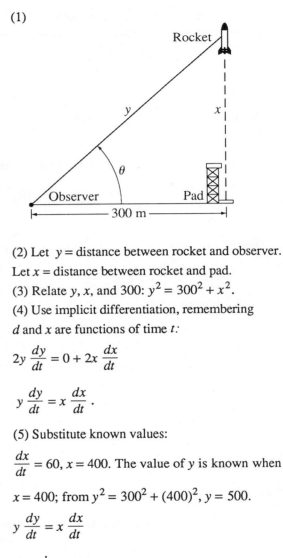

(2) Let $y =$ distance between rocket and observer.
Let $x =$ distance between rocket and pad.
(3) Relate y, x, and 300: $y^2 = 300^2 + x^2$.
(4) Use implicit differentiation, remembering
d and x are functions of time t:

$$2y \frac{dy}{dt} = 0 + 2x \frac{dx}{dt}$$

$$y \frac{dy}{dt} = x \frac{dx}{dt} \,.$$

(5) Substitute known values:

$\frac{dx}{dt} = 60$, $x = 400$. The value of y is known when

$x = 400$; from $y^2 = 300^2 + (400)^2$, $y = 500$.

$$y \frac{dy}{dt} = x \frac{dx}{dt}$$

$$500 \frac{dy}{dt} = 400(60)$$

$$\frac{dy}{dt} = 48 \text{ m/sec}.$$

2) A spherical ball has its diameter increasing at 2 inches/sec. How fast is the volume changing when the radius is 10 inches?

(1)

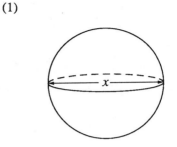

(2) Let x be the diameter, V the volume. We are given $\dfrac{dx}{dt} = 2$.

(3) Relate V and x:

$$V = \frac{4}{3}\pi \left(\frac{x}{2}\right)^3 = \frac{\pi}{6}x^3.$$

(4) Differentiate:

$$\frac{dV}{dt} = \frac{3\pi}{6}x^2\frac{dx}{dt} = \frac{\pi x^2}{2}\frac{dx}{dt}.$$

(5) When the radius is 10, $x = 20$.

$$\frac{dV}{dt} = \frac{\pi(20)^2}{2} \cdot 2 = 400\pi \text{ in}^3/\text{sec}.$$

3) A light is on top of a 12 ft vertical pole. A 6 ft woman walks away from the pole base at 4 ft/sec. How fast is the angle made by the woman's head, the light, and the pole base changing 2 seconds after she starts walking?

(1)

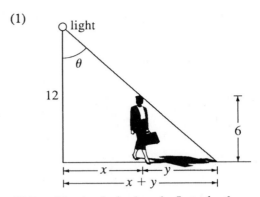

(2) Let θ be the desired angle. Let x be the distance between the pole base and the woman. Let y be the distance from the woman to the tip of her shadow.

(3) We are given $\dfrac{dx}{dt}$ ($= 4$) and we must find $\dfrac{d\theta}{dt}$.

Thus we must relate x and θ. We can relate $x + y$ and θ so we first determine y in terms of x. By similar triangles $\dfrac{x + y}{12} = \dfrac{y}{6}$.

Thus $6x + 6y = 12y$

$\qquad 6x = 6y$, or $x = y$.

Thus $x + y = x + x = 2x$.

By trigonometry, we obtain the desired relation between x and θ:

$\tan \theta = \dfrac{2x}{12}$, or $6 \tan \theta = x$.

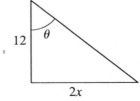

(4) Differentiate the relation:

$$6 \sec^2\theta \cdot \frac{d\theta}{dt} = \frac{dx}{dt}.$$

(5) After 2 seconds, $x = 2(4) = 8$ so $2x = 16$ ft.

$\sec \theta = \dfrac{d}{12}$, so $\sec^2\theta = \dfrac{d^2}{144}$.

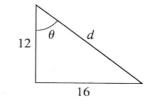

From the figure $12^2 + 16^2 = d^2$

so $d^2 = 400$ and $\sec^2\theta = \dfrac{400}{144}$.

Finally substituting into the result from (4):

$$6\left(\frac{400}{144}\right)\frac{d\theta}{dt} = 4$$

$$\frac{d\theta}{dt} = 2.4 \text{ radians/sec.}$$

Differentials and Linear Approximations

Concepts to Master

A. Differentials; Approximation with differentials
B. Linear approximation

Summary and Focus Questions

A. Let $y = f(x)$ be a differentiable function. The differential dx is an independent variable. The differential dy is defined as:

$$dy = f'(x)dx$$

Note that dy is a function of both x (because of $f'(x)$) and dx.

If we let $dx = \Delta x$, then for small values of dx, the change in the function (Δy) is approximately the same as the change in the tangent line dy:

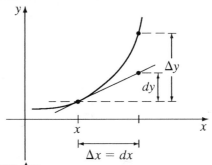

$$dy \approx \Delta y \text{ for } dx \text{ small.}$$

This is handy since dy may be easier to calculate than Δy.

1) True or False:
   $dy = \Delta y$.

   False.

2) Compute dy and Δy for $f(x) = x^2 + 3x$
   at $x = 2$ with $\Delta x = dx = .1$.

Section 2.9: Differentials and Linear Approximations

$f(x) = x^2 + 3x$

$f'(x) = 2x + 3.$

At $x = 2$, $f'(2) = 2(2) + 3 = 7.$

Thus $dy = f'(x)\, dx$

$\qquad = f'(2)(.1)$

$\qquad = 7(.1) = .7.$

At $x = 2$, $y = f(2) = 2^2 + 3(2) = 10.$

At $x = 2 + \Delta x = 2 + .1 = 2.1,$

$y = f(2.1) = (2.1)^2 + 3(2.1) = 10.71.$

Thus $\Delta y = f(2.1) - f(2)$

$\qquad = 10.71 - 10 = .71.$

3) A circle has a radius of 20 cm with a possible error of 0.2 cm. Use differentials to estimate the maximum error in the area of the circle.

Let x = radius of circle and A = area of circle.

We are given $\Delta x = dx = 0.2$ for $x = 20.$

The error is ΔA which we estimate with dA:

$A = \pi x^2$, so $dA = 2\pi x\, dx = 2\pi(20) \cdot (.02)$

$\qquad = .8\pi \ \text{cm}^2.$

B. Let $L(x)$ be the equation of the tangent line to $y = f(x)$ at $x = a$:

$$L(x) = f(a) + f'(a)\,(x - a)$$

For x near a, $L(x)$ may be used as a linear approximation to $f(x)$: $L(x) \approx f(x)$

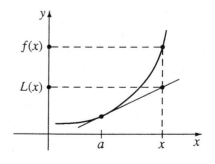

4) Find the linear approximation to $f(x) = 5x^3 + 6x$ at $x = 2.$

What we are asked to do is find the equation of the tangent line:

$f(2) = 5(2)^3 + 6(2) = 52.$

$f'(x) = 15x^2 + 6$, so
$f'(2) = 15(2)^2 + 6 = 66$.
Thus $L(x) = 52 + 66\,(x - 2)$.

5) Approximate $f(1.98)$ for the function in question 4.

We use $f(1.98) \approx L(1.98)$.
$L(1.98) = 52 + 66(1.98 - 2) = 50.68$.
(*Note*: $f(1.98) = 50.69196$, so $L(1.98)$ is rather close and a lot easier to calculate.)

6) Approximate $\sqrt{66}$ using a linear approximation.

Choose $f(x) = \sqrt{x}$ and $a = 64$
(64 is near 66). Then $f(64) = 8$,
$f'(x) = \dfrac{1}{2\sqrt{x}}$, and $f'(64) = \dfrac{1}{2\sqrt{64}} = \dfrac{1}{16}$.

The linear approximation is

$$L(x) = 8 + \frac{1}{16}(x - 64).$$

At $x = 66$, $L(66) = 8 + \dfrac{1}{16}(66 - 64) = 8.125$.

(*Note*: $\sqrt{66} \approx 8.1240384$).

7) Find the linear approximation of $f(x) = \sin x$ at $a = 0$.

$f(x) = \sin x$
$f(0) = \sin 0 = 0$.
$f'(x) = \cos x$
$f'(0) = \cos 0 = 1$.
So $L(x) = f(0) + f'(0)\,(x - 0) =$
$0 + 1(x - 0) = x$.
Thus $\sin x \approx x$, for x near 0.
(*Note*: This is an approximation that you may see in other courses such as physics.)

Newton's Method

Concepts to Master

Approximation of solutions to $f(x) = 0$ using Newton's Method

Summary and Focus Questions

Let f be a continuously differentiable function on an open interval with a real root. If x_1 is an estimate of the root, then $x_2 = x_1 - \dfrac{f(x_1)}{f'(x_1)}$ is

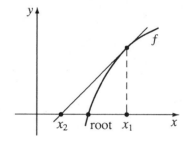

often (but not always) a closer approximation to the root. The point x_2 is where the tangent to f at x_1 crosses the x-axis. The process of obtaining x_2 may be repeated, using x_2 in the role of x_1 to obtain a sequence of approximations $x_1, x_2, x_3, \ldots$ which can approach the value of the root. In general, to obtain a desired degree of accuracy, repeat Newton's Method until the difference between successive x_i is within that accuracy.

1) Indicate the point x_2 determined by Newton's Method in each:

   a)

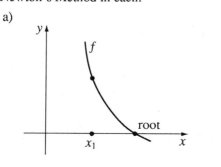

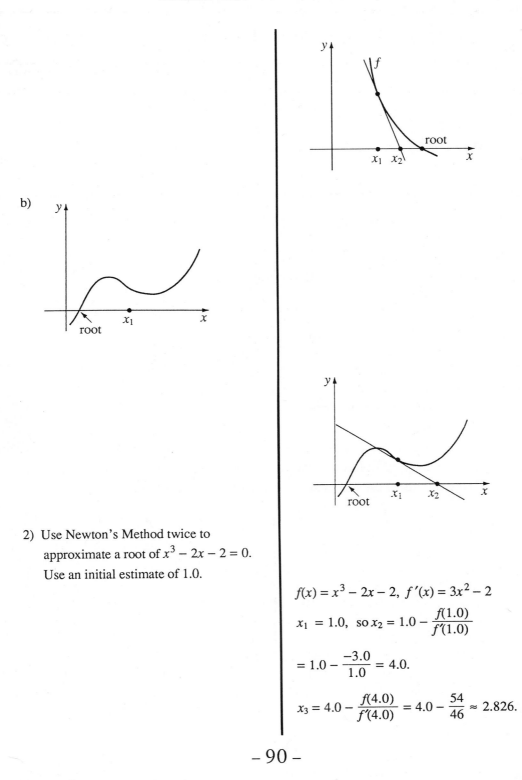

b)

2) Use Newton's Method twice to approximate a root of $x^3 - 2x - 2 = 0$. Use an initial estimate of 1.0.

$f(x) = x^3 - 2x - 2,\ f'(x) = 3x^2 - 2$

$x_1 = 1.0,\ \ \text{so } x_2 = 1.0 - \dfrac{f(1.0)}{f'(1.0)}$

$= 1.0 - \dfrac{-3.0}{1.0} = 4.0.$

$x_3 = 4.0 - \dfrac{f(4.0)}{f'(4.0)} = 4.0 - \dfrac{54}{46} \approx 2.826.$

3

Inverse Functions: Exponential, Logarithmic, and Inverse Trigonometric Functions

"IF IT'S TRUE THAT THE WORLD ANT POPULATION IS 10^{15} THEN IT'S NO WONDER WE NEVER RUN INTO ANYONE WE KNOW."

Exponential Functions

Concepts to Master

Definition, properties and graph of an exponential function

Summary and Focus Questions

Let a be a positive constant. The exponential function with base a is $f(x) = a^x$ and is defined as follows:

Type of value of x	Definition of a^x
n, a positive integer	$a^n = a \cdot a \cdot a \ldots$ (n times)
0	$a^0 = 1$
$-n$ (n, a positive integer)	$a^{-n} = \dfrac{1}{a^n}$
$\dfrac{p}{q}$, a rational number	$a^{p/q} = \sqrt[q]{a^p}$
x, irrational	$a^x = \lim\limits_{r \to x} a^r$, where r is rational

$f(x) = a^x$ is continuous for all real x. The graph of $y = a^x$ is:

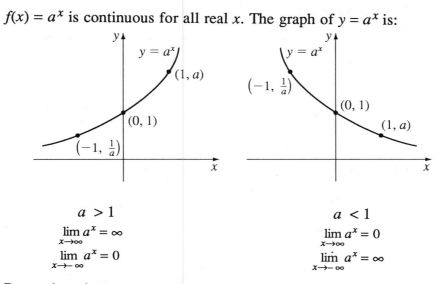

$$a > 1$$
$$\lim_{x \to \infty} a^x = \infty$$
$$\lim_{x \to -\infty} a^x = 0$$

$$a < 1$$
$$\lim_{x \to \infty} a^x = 0$$
$$\lim_{x \to -\infty} a^x = \infty$$

Properties of exponentials include:

$$a^{x+y} = a^x a^y$$
$$a^{x-y} = \frac{a^x}{a^y}$$
$$a^{xy} = (a^x)^y$$
$$(ab)^x = a^x b^x$$

1) By definition, $2^{4/3} =$ _____ .

$$\sqrt[3]{2^4}$$

2) $\lim_{r \to \sqrt{3}} 3^r =$ _____ , where r is rational.

$$3^{\sqrt{3}}$$

3) $\lim_{x \to -\infty} (\frac{1}{2})^x =$ _____ .

$$\infty$$

4) Graph $y = (0.7)^x$.

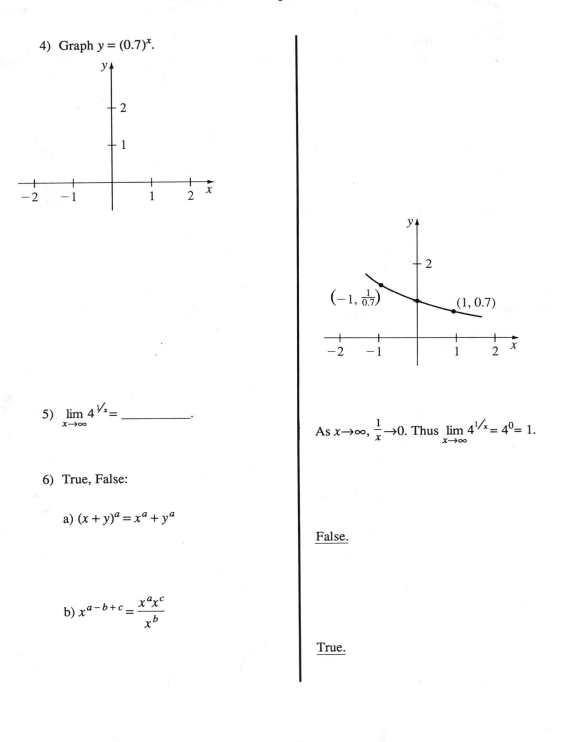

$\left(-1, \frac{1}{0.7}\right)$ $(1, 0.7)$

5) $\lim\limits_{x \to \infty} 4^{1/x} =$ _____ .

As $x \to \infty$, $\frac{1}{x} \to 0$. Thus $\lim\limits_{x \to \infty} 4^{1/x} = 4^0 = 1$.

6) True, False:

a) $(x + y)^a = x^a + y^a$

False.

b) $x^{a-b+c} = \dfrac{x^a x^c}{x^b}$

True.

Derivatives of Exponential Functions

Concepts to Master

A. Definition of e; Properties of $y = e^x$
B. Derivative of e^x

Summary and Focus Questions

A. The number e is a constant (approximately 2.718) such that $\lim\limits_{h \to 0} \dfrac{e^h - 1}{h} = 1$.

The function $f(x) = e^x$ has the usual exponential function properties: continuous, increasing, concave upward,

$$\lim_{x \to \infty} e^x = \infty,$$

$$\lim_{x \to -\infty} e^x = 0.$$

(graph showing $y = e^x$ with points $(1, e)$, $(1, 0)$, and $\left(-1, \frac{1}{e}\right)$)

1) For what number(s) k is $\lim\limits_{h \to 0} \dfrac{k^h - 1}{h} = 1$?

$k = e$ is the only such number.

2) $\lim\limits_{x \to \infty} e^{-x}$

As $x \to \infty$, $-x \to -\infty$.
Thus $\lim\limits_{x \to \infty} e^{-x} = 0$.

Section 3.2: Derivatives of Exponential Functions

B. In general, $(a^x)' = a^x \cdot \left(\lim\limits_{h \to 0} \dfrac{a^h - 1}{h}\right)$. Thus the number e was chosen in order to make this differentiation formula simple:

$$\text{If } f(x) = e^x, \text{ then } f'(x) = e^x.$$

$$\frac{d\, e^{u(x)}}{dx} = e^{u(x)} u'(x).$$

For example, if $f(x) = e^{3x^2}$, then $f'(x) = e^{3x^2}(6x)$.

3) Find $f'(x)$ for:

a) $f(x) = e^{\sqrt{x}}$

$$f(x) = e^{x^{\frac{1}{2}}}, \quad f'(x) = e^{x^{\frac{1}{2}}}\left(\frac{1}{2}x^{-\frac{1}{2}}\right)$$
$$= \frac{e^{\sqrt{x}}}{2\sqrt{x}}.$$

b) $f(x) = e^3$

$f'(x) = 0$. (Remember, e is a constant and therefore e^3 is a constant).

c) $f(x) = x^2 \cos e^x$

$$f'(x) = x^2\,(-\sin e^x\, e^x) + \cos e^x(2x)$$
$$= -x^2\, e^x \sin e^x + 2x \cos e^x.$$

4) For what a is $(a^x)' = a^x$?

Only for $a = e$.

5) Simplify to e to a power:

$$\frac{(e^x \cdot e^3)^2}{e}$$

$$\frac{(e^x \cdot e^3)^2}{e} = \frac{(e^{x+3})^2}{e} = \frac{e^{2x+6}}{e}$$
$$= e^{2x+6-1} = e^{2x+5}.$$

Inverse Functions

Concepts to Master

A. One-to-one functions; Inverse of a function
B. Derivative of the inverse function

Summary and Focus Questions

A. A function f is <u>one-to-one</u> means that for all x_1, x_2 in the domain, if $x_1 \neq x_2$, then $f(x_1) \neq f(x_2)$.

For example, $f(x) = 2x + 5$ is one-to-one because if $x_1 \neq x_2$, $2x_1 + 5 \neq 2x_2 + 5$, while $f(x) = x^2$ is not one-to-one ($4 \neq -4$, yet $f(4) = 4^2 = (-4)^2 = f(-4)$).

If f is one-to-one with domain A and range B then the inverse function f^{-1} has domain B, range A, and
$$f^{-1}(b) = a \quad \text{iff} \quad f(a) = b.$$

Thus a point (a, b) is on the graph of f if and only if (b, a) is on the graph of f^{-1}. To find the rule for $y = f^{-1}(x)$,

(1) solve the equation $y = f(x)$ for x.
(2) interchange x and y.

1) Is $f(x) = \sin x$ one-to-one?

No. For example, $0 \neq \pi$,
yet $\sin 0 = 0 = \sin \pi$.

2) Given the following graph of f, sketch the graph of f^{-1} on the same axis.

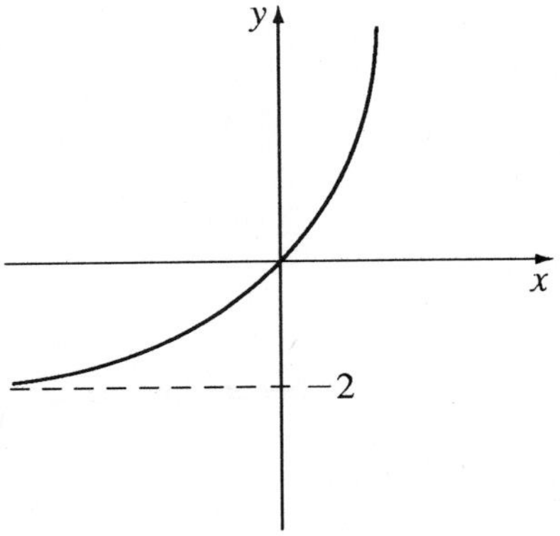

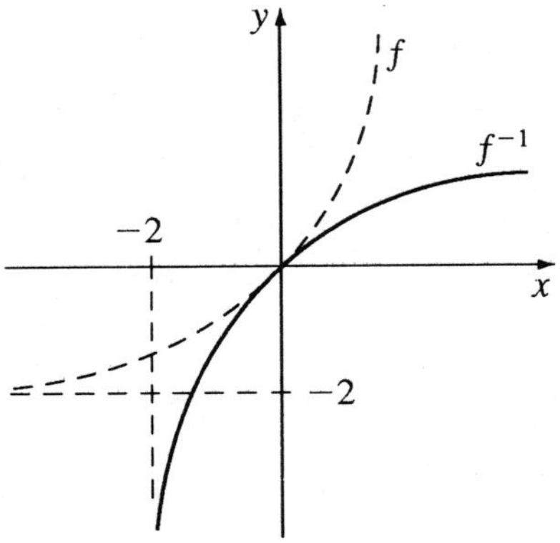

3) If $f(5) = 8$, then $f^{-1}(8) =$ _____.

4) Write the inverse function of
$f(x) = 6x + 30$.

5.

(1) $y = 6x + 30$

$y - 30 = 6x$

$x = \dfrac{1}{6}y - 5$

(2) $y = \dfrac{1}{6}x - 5$, so $f^{-1}(x) = \dfrac{1}{6}x - 5$.

B. If f is one-to-one, differentiable, and $f(b) = a$, then $(f^{-1})'(a) = \dfrac{1}{f'(b)}$, provided

$f'(b) \neq 0$.

Since $f^{-1}(a) = (b)$, this may also be written as

$$(f^{-1})'(a) = \frac{1}{f'(f^{-1}(a))}.$$

5) Suppose $f(3) = 5, f'(3) = 6$ and f is
   one-to-one.
   Then $(f^{-1}(5))' = $ _____ .

$$\frac{1}{f'(3)} = \frac{1}{6}.$$

6) Let $f(x) = x^3 + 2x + 3$. Find $(f^{-1})'(3)$.

We first must find b such that $b = f^{-1}(3)$;
that is, $f(b) = 3$.
$f(b) = b^3 + 2b + 3 = 3$
$b^3 + 2b = 0$
$b(b^2 + 2) = 0$.
Since $b^2 + 2 \neq 0$, $b = 0$.
$f'(x) = 3x^2 + 2$, so
$f'(0) = 3(0^2) + 2 = 2$.
Hence $(f^{-1})'(3) = \dfrac{1}{f'(0)} = \dfrac{1}{2}$.

Logarithmic Functions

Concepts to Master

Definition, properties, and graph of $y = \log_a x$.

Summary and Focus Questions

$\log_a x = y$ means $a^y = x$.
This definition requires $x > 0$, since $a^y > 0$ for all y. $\log_a x$ is the exponent you put on a to get x.

The logarithmic function $y = \log_a x$ is defined as the inverse of the exponential $y = a^x$. Thus

$$\log_a (a^x) = x$$

$$a^{\log_a x} = x, \text{ for } x > 0$$

$$\log_a (xy) = \log_a x + \log_a y$$

$$\log_a \left(\frac{x}{y}\right) = \log_a x - \log_a y$$

$$\log_a x^c = c \log_a x$$

$$\lim_{x \to \infty} \log_a x = \infty$$

$$\lim_{x \to 0^+} \log_a x = -\infty$$

For $a > 1$, the graph of $y = \log_a x$ is

$y = \log_a x$ $(a, 1)$

$(1, 0)$

$\left(\frac{1}{a}, -1\right)$

Section 3.4: Logarithmic Functions

In the special case that $a = e$, <u>$\ln x$</u> means $\log_a x$.

1) $\log_4 8 = $ _____ .

Let $y = \log_4 8$.
Then $4^y = 8$
$2^{2y} = 2^3$
$2y = 3$
$y = \dfrac{2}{3}$.

2) True or False:

   a) $\log_a x^3 = 3 \log_a x$

   <u>True.</u>

   b) $\log_2 (10) = \log_2(-5) + \log_2(-2)$.

   <u>False.</u>

   c) $\ln(a - b) = \ln a - \ln b$.

   <u>False.</u>

   d) $\ln e = 1$

   <u>True.</u> In $\log_e$ notation this is $\log_e e = 1$.

   e) $\ln\left(\dfrac{a}{b}\right) = -\ln\left(\dfrac{b}{a}\right)$

   <u>True.</u>

3) Write $2 \ln x + 3 \ln y$ as a single logarithm.

$2 \ln x + 3 \ln y = \ln x^2 + \ln y^3$
$= \ln (x^2 y^3)$.

Section 3.4: Logarithmic Functions

4) For $x > 0$, simplify using properties of $\log_2$:

$$\log_2\left(\frac{4x^3}{\sqrt{2}}\right)$$

$$\log_2\left(\frac{4x^3}{\sqrt{2}}\right) = \log_2 4x^3 - \log_2 \sqrt{2}$$

$$= \log_2 4 + \log_2 x^3 - \log_2 2^{\frac{1}{2}}$$

$$= \log_2 4 + 3\log_2 x - \frac{1}{2}\log_2 2$$

$$= 2 + 3\log_2 x - \frac{1}{2}(1)$$

$$= \frac{3}{2} + 3\log_2 x.$$

5) Solve for x:

   a) $e^{x+1} = 10$

   By definition, this means $\ln 10 = x + 1$, so $x = \ln 10 - 1$.

   b) $\ln(x + 5) = 2$

   By definition, this means $e^2 = x + 5$, so $x = e^2 - 5$.

   c) $2^{3x} = 7$

   Take $\log_2$ of both sides:
   $\log_2 2^{3x} = \log_2 7$
   $3x = \log_2 7$
   $x = \frac{1}{3}\log_2 7.$

6) $\lim\limits_{x \to 0} \ln(\cos x)$

   As $x \to 0$, $\cos x \to 1$.
   Thus $\ln(\cos x) \to \ln(1) = 0$.

Derivatives of Logarithmic Functions

Concepts to Master

A.   Derivatives of ln x, $\log_a x$, and a^x
B.   Logarithmic differentiation
C.   The number e as a limiting value

Summary and Focus Questions

A. If $f(x) = \ln x$, $f'(x) = \dfrac{1}{x}$.

If $f(x) = \log_a x$, $f'(x) = \dfrac{1}{x \ln a}$.

If $f(x) = a^x$, $f'(x) = a^x \ln a$.

1) Find $f'(x)$ for each:

a) $f(x) = \ln x^2$

$f(x) = 2 \ln x$, so $f'(x) = 2\left(\dfrac{1}{x}\right) = \dfrac{2}{x}$.

Alternatively, using the Chain Rule,

$f'(x) = \dfrac{1}{x^2}\,(2x) = \dfrac{2}{x}$.

b) $f(x) = \ln \cos x$

$$f'(x) = \frac{1}{\cos x} \cdot (-\sin x) = -\tan x.$$

c) $f(x) = \log_2 (x^2 + 1)$

$$f'(x) = \frac{1}{\ln 2\,(x^2 + 1)} \cdot (2x) = \frac{2x}{\ln 2\,(x^2 + 1)}.$$

d) $f(x) = 4^x$

$$f'(x) = 4^x \ln 4.$$

e) $f(x) = 2^x(x^2 + 1)$

$$f'(x) = 2^x(2x) + (x^2 + 1)\,2^x \ln 2$$
$$= 2^x(2x + \ln 2\,(x^2 + 1)).$$

f) $f(x) = \ln \sqrt{\dfrac{x^2 + 1}{2x^3}} \quad (x > 0)$

You should simplify y first:

$$f(x) = \frac{1}{2} \ln \left(\frac{x^2 + 1}{2x^3} \right)$$

$$= \frac{1}{2} (\ln (x^2 + 1) - \ln 2 - 3 \ln x)$$

$$f'(x) = \frac{1}{2} \left(\frac{2x}{x^2 + 1} - \frac{3}{x} \right).$$

Section 3.5: Derivatives of Logarithmic Functions

B. The derivative of a function $y = f(x)$ may be found using logarithmic differentiation, that is,

$$\frac{dy}{dx} = y \frac{d}{dx}(\ln y).$$

2) Find y' for each, using logarithmic differentiation.

a) $y = \left(\dfrac{10x^3}{\sqrt{x+1}}\right)^4$

$$\ln y = \ln \left(\frac{10x^3}{\sqrt{x+1}}\right)^4$$

$$= 4\left(\ln 10x^3 - \ln\sqrt{x+1}\right)$$

$$= 4\left(\ln 10 + 3\ln x - \frac{1}{2}\ln(x+1)\right)$$

$$\frac{d}{dx}\ln y = 4\left(0 + \frac{3}{x} - \frac{1}{2}\frac{1}{x+1}(1)\right)$$

$$= 4\left(\frac{3}{x} - \frac{1}{2(x+1)}\right).$$

Thus $y' = \left(\dfrac{10x^3}{\sqrt{x+1}}\right)^4\left[4\left(\dfrac{3}{x} - \dfrac{1}{2(x+1)}\right)\right].$

b) $y = 6^x$

$$\ln y = \ln 6^x$$
$$= x \ln 6.$$

Thus $\dfrac{d}{dx}\ln y = \ln 6$ (a constant)

and $y' = y \ln 6 = 6^x \ln 6.$

– 105 –

C. $\lim\limits_{x \to 0^+} (1 + x)^{\frac{1}{x}} = e$. Equivalently, $\lim\limits_{t \to \infty} \left(1 + \dfrac{1}{t}\right)^t = e$.

3) Find each limit:

   a) $\lim\limits_{h \to 0^+} (1 + h)^{1/h}$

                                                e

   b) $\lim\limits_{x \to \infty} \left(1 + \dfrac{1}{x}\right)^{3x}$

$$\lim\limits_{x \to \infty} \left(1 + \frac{1}{x}\right)^{3x} = \lim\limits_{x \to \infty} \left[\left(1 + \frac{1}{x}\right)^{x}\right]^3$$

$$= \left[\lim\limits_{x \to \infty} \left(1 + \frac{1}{x}\right)^{x}\right]^3 = e^3.$$

   c) $\lim\limits_{x \to 1^+} x^{\frac{1}{x-1}}$

Let $t = x - 1$, hence $x = 1 + t$.
As $x \to 1^+$, $t \to 0^+$. Thus $\lim\limits_{x \to 1^+} x^{\frac{1}{x-1}}$

$$= \lim\limits_{t \to 0^+} (1 + t)^{1/t} = e.$$

Exponential Growth and Decay

Concepts to Master

Solution to $y' = ky$; Applications to growth and decay.

Summary and Focus Questions

The solution to the differential equation $\dfrac{dy}{dt} = ky$ is $y(t) = y(0) \cdot e^{kt}$.

This has applications where the quantity (y) of a substance varies with time such that the rate of change (y') of the quantity is proportional to y — that is, $y' = ky$. If $k > 0$ this is <u>natural growth</u>; if $k < 0$, this is <u>natural decay</u>.

Many growth and decay problems involving $y(t) = y(0)e^{kt}$ boil down to: "given $y = y_1$ at time t_1 and $y = y_2$ at time t_2, find y at some third time t_3." The procedure for solving $y' = ky$ depends on whether or not one of the times given is $t = 0$.

If $t_1 = 0$, then we are given $y(0)$. Then just solve $y_2 = y(0) \cdot e^{kt_2}$ for k.

If neither t_1 nor t_2 are zero, solve the equations $y_1 = y(0) \cdot e^{kt_1}$ and $y_2 = y(0) \cdot e^{kt_2}$ simultaneously for $y(0)$ and k. You first divide one equation by the other to eliminate the $y(0)$ term, solving for k.

1) A village had a population of 1000 in 1980 and 1200 in 1990. Assuming the population is experiencing natural growth, what will be the population in 2010 ?

Let 1980 be $t = 0$. Then $y(0) = 1000$.

At $t_1 = 10$ (year 1990), $y = 1200$

so $1200 = 1000 \, e^{k(10)}$

$1.2 = e^{10k}$

$10k = \ln 1.2$

$k = \dfrac{1}{10} \ln 1.2$.

Thus $y = 1000 \, e^{(1/10 \, \ln 1.2)t}$

$= 1000 \, (e^{\ln 1.2})^{t/10}$

$= 1000 \, (1.2)^{t/10}$.

In the year 2010, $t = 30$, so

$y = 1000(1.2)^{30/10} = 1000 \, (1.2)^3$.

2) One hour after a bacteria culture was started there were 300 organisms and after 2 hours from the start there were 900. How many organisms will there be 4 hours after the start?

At $t = 1$, $y = 300$; $t = 2$, $y = 900$.

Thus $300 = y(0) \, e^{k \cdot 1}$

$900 = y(0) \, e^{k \cdot 2}$.

Dividing the second equation by the first

$\dfrac{900}{300} = \dfrac{y(0) \, e^{2k}}{y(0) \, e^k}$

$3 = e^k$

$k = \ln 3$.

Now use $300 = y(0) \, e^{k \cdot 1}$ to solve for $y(0)$

$300 = y(0) \, e^{\ln 3}$

$300 = y(0) \, 3$

$y(0) = 100$.

The model is $y = 100 \, e^{\ln 3 \, t} = 100 \, (e^{\ln 3})^t = 100 \cdot 3^t$.

So $y = 100 \cdot 3^t$.

At $t = 4$, $y = 100 \cdot 3^4 = 8100$.

Inverse Trigonometric Functions

Concepts to Master

A. Definitions and functional values of the six inverse trigonometric functions
B. Derivatives involving inverse trigonometric functions

Summary and Focus Questions

A. For any function f for which f^{-1} is a function, $y = f^{-1}(x)$ iff $x = f(y)$.

When the domains of the trigonometric function are properly restricted they have inverse functions. For example, sine with domain $[-\pi/2, \pi/2]$ has an inverse function $\sin^{-1}$. Thus $y = \sin^{-1}\left(\dfrac{1}{2}\right)$ means $\sin(y) = \dfrac{1}{2}$ and $-\pi/2 \le y \le \pi/2$; so $y = \pi/6$.

The domains and ranges of the six inverse trigonometric functions are:

Function	Domain	Range
$\sin^{-1}$	$[-1, 1]$	$[-\pi/2, \pi/2]$
$\cos^{-1}$	$[-1, 1]$	$[0, \pi]$
$\tan^{-1}$	all reals	$(-\pi/2, \pi/2)$
$\cot^{-1}$	all reals	$(0, \pi)$
$\sec^{-1}$	$(-\infty, -1] \cup [1, \infty)$	$[0, \pi/2) \cup [\pi, 3\pi/2)$
$\csc^{-1}$	$(-\infty, -1] \cup [1, \infty)$	$(0, \pi/2] \cup (\pi, 3\pi/2]$

The first three functions are used frequently; their graphs are:

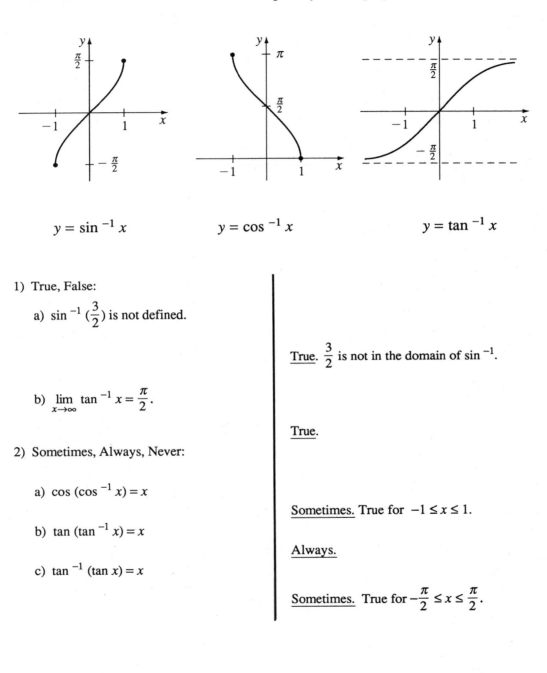

$$y = \sin^{-1} x \qquad\qquad y = \cos^{-1} x \qquad\qquad\qquad y = \tan^{-1} x$$

1) True, False:

a) $\sin^{-1}(\frac{3}{2})$ is not defined.

<u>True.</u> $\frac{3}{2}$ is not in the domain of $\sin^{-1}$.

b) $\lim\limits_{x\to\infty} \tan^{-1} x = \frac{\pi}{2}$.

<u>True.</u>

2) Sometimes, Always, Never:

a) $\cos(\cos^{-1} x) = x$

<u>Sometimes.</u> True for $-1 \le x \le 1$.

b) $\tan(\tan^{-1} x) = x$

<u>Always.</u>

c) $\tan^{-1}(\tan x) = x$

<u>Sometimes.</u> True for $-\frac{\pi}{2} \le x \le \frac{\pi}{2}$.

3) Determine each:

   a) $\cos^{-1}\left(\frac{1}{2}\right)$

Let $y = \cos^{-1}\left(\frac{1}{2}\right)$.

Then $0 \le y \le \pi$ and $\frac{1}{2} = \cos(y)$.

From your knowledge of trigonometry, $y = \frac{\pi}{3}$.

   b) $\tan^{-1}(\sqrt{3})$

Let $y = \tan^{-1}\left(-\sqrt{3}\right)$. Then $-\frac{\pi}{2} < y < \frac{\pi}{2}$

and $-\sqrt{3} = \tan y$.

Thus $y = -\frac{\pi}{3}$.

   c) $\sin^{-1}\left(\sin \frac{3\pi}{4}\right)$

Since $\sin \frac{3\pi}{4} = \frac{1}{\sqrt{2}}$, this problem asks you to find

$y = \sin^{-1}\left(\frac{1}{\sqrt{2}}\right)$. Then $-\frac{\pi}{2} \le y \le \frac{\pi}{2}$ and

$\sin y = \frac{1}{\sqrt{2}}$.

Thus $y = \frac{\pi}{4}$.

   d) $\sin^{-1}\left(\sin \frac{\pi}{3}\right)$

For $-\frac{\pi}{2} \le x \le \frac{\pi}{2}$, $\sin^{-1}(\sin x) = x$.

Thus $\sin^{-1}\left(\sin \frac{\pi}{3}\right) = \frac{\pi}{3}$.

e) $\sec^{-1}\left(\sin\dfrac{\pi}{7}\right)$

This does not exist because $-1 \le \sin\dfrac{\pi}{7} \le 1$, so

$\sin\dfrac{\pi}{7}$ is not in the domain of $\sec^{-1}$.

4) For $0 < x < 1$, simplify $\tan(\cos^{-1}x)$

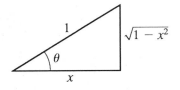

If $\cos\theta = x$ where $0 < \theta < \dfrac{\pi}{2}$ then $\theta = \cos^{-1}x$

and $\tan\theta = \dfrac{\sqrt{1-x^2}}{x}$.

B. Here are the derivatives of the inverse trigonometric functions

$$\frac{d}{dx}\sin^{-1}x = \frac{1}{\sqrt{1-x^2}} \qquad \frac{d}{dx}\cos^{-1}x = \frac{-1}{\sqrt{1-x^2}}$$

$$\frac{d}{dx}\tan^{-1}x = \frac{1}{1+x^2} \qquad \frac{d}{dx}\cot^{-1}x = \frac{-1}{1+x^2}$$

$$\frac{d}{dx}\sec^{-1}x = \frac{1}{x\sqrt{x^2-1}} \qquad \frac{d}{dx}\csc^{-1}x = \frac{-1}{x\sqrt{x^2-1}}$$

5) Find $f'(x)$ for each:

   a) $f(x) = \sin^{-1}(x^2)$

$$f'(x) = \frac{1}{\sqrt{1-(x^2)^2}}\,(2x) = \frac{2x}{\sqrt{1-x^4}}.$$

   b) $f(x) = (\tan^{-1}x)^3$

$$f'(x) = 3\,(\tan^{-1}x)^2 \cdot \frac{1}{1+x^2}$$

$$= \frac{3\,(\tan^{-1}x)^2}{1+x^2}.$$

   c) $f(x) = x\,\cos^{-1}x$

$$f'(x) = x\left(\frac{-1}{\sqrt{1-x^2}}\right) + \cos^{-1}x\,(1)$$

$$= \frac{-x}{\sqrt{1-x^2}} + \cos^{-1}x.$$

6) Find y' for $y = \sin^{-1}x + \cos^{-1}x$.
   What can you conclude about
   $\sin^{-1}x + \cos^{-1}x$?

$$y' = \frac{1}{\sqrt{1-x^2}} + \frac{-1}{\sqrt{1-x^2}} = 0.$$

Thus y is a constant. To determine what constant, choose $x = 0$,

$$y = \sin^{-1}0 + \cos^{-1}0 = 0 + \frac{\pi}{2} = \frac{\pi}{2}.$$

Therefore $\sin^{-1}x + \cos^{-1}x = \dfrac{\pi}{2}$.

Hyperbolic Functions

Concepts to Master

A. Define the hyperbolic trigonometric functions
B. Derivatives involving hyperbolic functions
C. Definition and derivatives of inverse hyperbolic functions

Summary and Focus Questions

A. The hyperbolic sine and cosine functions are

$$\sinh x = \frac{e^x - e^{-x}}{2} \qquad \cosh x = \frac{e^x + e^{-x}}{2}$$

The other hyperbolic trigonometric functions are

$$\tanh x = \frac{\sinh x}{\cosh x} \qquad \coth x = \frac{\cosh x}{\sinh x}$$

$$\text{sech } x = \frac{1}{\cosh x} \qquad \text{csch } x = \frac{1}{\sinh x}$$

One reason why they are called hyperbolic is that $\cosh^2 t - \sinh^2 t = 1$ and so $(\cosh t, \sinh t)$ is on the hyperbola $x^2 - y^2 = 1$. (Recall the trigonometric functions are sometimes called "circular" functions because $(\cos t, \sin t)$ is on the circle $x^2 + y^2 = 1$.)

1) Evaluate sinh 2.

$$\sinh 2 = \frac{e^2 - e^{-2}}{2} = \frac{e^4 - 1}{2e^2}.$$

Section 3.8: Hyperbolic Functions

2) Write $\tanh x$ in terms of e^x.

$$\tanh x = \frac{\cosh x}{\sinh x} = \frac{\dfrac{e^x + e^{-x}}{2}}{\dfrac{e^x - e^{-x}}{2}} = \frac{e^x + e^{-x}}{e^x - e^{-x}}.$$

3) Verify $\sinh 2x = 2 \sinh x \cosh x$.

$$2 \sinh x \cosh x = 2 \, \frac{e^x - e^{-x}}{2} \cdot \frac{e^x + e^{-x}}{2}$$

$$= \frac{(e^x - e^{-x})(e^x + e^{-x})}{2}$$

$$= \frac{e^{2x} - e^{-2x}}{2} = \sinh 2x.$$

B. Here are derivatives for the hyperbolics:

$$\frac{d}{dx} \sinh x = \cosh x \qquad\qquad \frac{d}{dx} \cosh x = \sinh x$$

$$\frac{d}{dx} \tanh x = \operatorname{sech}^2 x \qquad\qquad \frac{d}{dx} \coth x = - \operatorname{csch}^2 x$$

$$\frac{d}{dx} \operatorname{sech} x = - \operatorname{sech} x \tanh x \qquad\qquad \frac{d}{dx} \operatorname{csch} x = - \operatorname{csch} x \coth x$$

4) Find $f'(x)$ for each:

a) $f(x) = \tanh\left(\dfrac{x}{2}\right)$

$$f'(x) = \operatorname{sech}^2\left(\frac{x}{2}\right)\left(\frac{1}{2}\right)$$

$$= \frac{1}{2} \operatorname{sech}^2\left(\frac{x}{2}\right).$$

b) $f(x) = \sqrt{\cosh x}$

$$f'(x) = \frac{1}{2} (\cosh)^{-1/2} (\sinh x)$$

$$= \frac{\sinh x}{2\sqrt{\cosh x}} .$$

C. The inverses of the hyperbolic functions are:

Function	Domain	Range	Explicit Rule
$\sinh^{-1} x$	all reals	all reals	$\ln\left(x + \sqrt{x^2 + 1}\right)$
$\cosh^{-1} x$	$[1, \infty)$	$[0, \infty)$	$\ln\left(x + \sqrt{x^2 - 1}\right)$
$\tanh^{-1} x$	$(-1, 1)$	all reals	$\frac{1}{2} \ln\left(\frac{1 + x}{1 - x}\right)$
$\coth^{-1} x$	$(-\infty, -1) \cup (1, \infty)$	all reals except 0	$\frac{1}{2} \ln\left(\frac{1 + x}{x - 1}\right)$
$\text{sech}^{-1} x$	$(0, 1]$	$[0, \infty)$	$\ln\left(\frac{1}{x} + \sqrt{\frac{1}{x^2} - 1}\right)$
$\text{csch}^{-1} x$	all reals except 0	all reals except 0	$\ln\left(\frac{1}{x} + \sqrt{\frac{1}{x^2} + 1}\right)$

The derivatives are:

$$\frac{d}{dx} \sinh^{-1} x = \frac{1}{\sqrt{1 + x^2}}$$

$$\frac{d}{dx} \cosh^{-1} x = \frac{1}{\sqrt{x^2 - 1}}$$

$$\frac{d}{dx} \tanh^{-1} x = \frac{1}{1 - x^2}$$

$$\frac{d}{dx} \text{csch}^{-1} x = \frac{-1}{|x| \sqrt{x^2 + 1}}$$

$$\frac{d}{dx} \text{sech}^{-1} x = \frac{-1}{x \sqrt{1 - x^2}}$$

$$\frac{d}{dx} \coth^{-1} x = \frac{1}{1 - x^2}$$

5) Show that $f(x) = \tanh^{-1}x$ is always increasing.

$$f'(x) = \frac{1}{1 - x^2}$$

Since the domain is $(-1, 1)$ we have

$-1 < x < 1$, hence $\dfrac{1}{1 - x^2} > 0.$

6) Verify the derivative $\dfrac{d}{dx} \coth^{-1}x = \dfrac{1}{1 - x^2}$, using the explicit rule for $\coth^{-1} x$.

$$\coth^{-1}x = \frac{1}{2} \ln\left(\frac{1 + x}{x - 1}\right).$$

Therefore $\dfrac{d}{dx} \coth^{-1}x$

$$= \frac{1}{2} \; \frac{1}{\dfrac{1 + x}{x - 1}} \; \frac{(x - 1)(1) - (1 + x)(1)}{(x - 1)^2}$$

$$= \frac{1}{2} \; \frac{x - 1}{1 + x} \; \frac{-2}{(x - 1)^2}$$

$$= \frac{-1}{(1 + x)(x - 1)} = \frac{-1}{x^2 - 1} = \frac{1}{1 - x^2}.$$

7) Find $f'(x)$ for $f(x) = \sinh^{-1}x^2.$

$$f'(x) = \frac{1}{\sqrt{1 + (x^2)^2}} \, (2x)$$

$$= \frac{2x}{\sqrt{1 + x^4}}.$$

Indeterminate Forms and L'Hospital's Rule

Concepts to Master

Recognition of indeterminate forms; L'Hospital's Rule.

Summary and Focus Questions

Sometimes applying the limit theorems of Chapter 1 results in a nonsensical expression yet the limit still exists. Several of these expressions are called indeterminate forms. For example, $\lim_{x \to 2} \dfrac{x^2 - 4}{x - 2}$ has the indeterminate form $\dfrac{0}{0}$ but the limit is 4.

Forms $\dfrac{0}{0}$ and $\dfrac{\infty}{\infty}$:

Limits of the form $\lim_{x \to a} \dfrac{f(x)}{g(x)}$ which have indeterminate form $\dfrac{0}{0}$ or $\dfrac{\infty}{\infty}$ are amenable to L'Hospital's Rule:

$$\lim_{x \to a} \frac{f(x)}{g(x)} = \lim_{x \to a} \frac{f'(x)}{g'(x)}.$$

Example:

Although $\lim_{x \to 2} \dfrac{x^2 - 4}{x - 2}$ can be evaluated by canceling first, L'Hospital's Rule is faster:

$$\lim_{x \to 2} \frac{x^2 - 4}{x - 2} = \lim_{x \to 2} \frac{2x}{1} = 4.$$

Section 3.9: Indeterminate Forms and L'Hospital's Rule

Forms $0 \cdot \infty$:

To evaluate $\lim\limits_{x \to a} f(x) \cdot g(x)$ which has indeterminate form $0 \cdot \infty$, first rewrite $f(x) \cdot g(x)$ as either $\dfrac{f(x)}{1/g(x)}$ or $\dfrac{g(x)}{1/f(x)}$ to get either $\dfrac{0}{0}$ or $\dfrac{\infty}{\infty}$. Now apply L'Hospital's Rule.

Example: $\lim\limits_{x \to \infty} x\, e^{-x}$ (form $\infty \cdot 0$) $= \lim\limits_{x \to \infty} \dfrac{x}{e^x}$ $\left(\text{form } \dfrac{\infty}{\infty}\right) = \lim\limits_{x \to \infty} \dfrac{1}{e^x} = 0.$

Form $\infty - \infty$:

For $\lim\limits_{x \to a} f(x) - g(x) = \infty - \infty$, $f(x) - g(x)$ must be rewritten as a single term. When the trigonometric functions are involved, switching to all sines and cosines may help.

Example: $\lim\limits_{x \to \frac{\pi}{2}^-} (\sec x - \tan x) = \lim\limits_{x \to \frac{\pi}{2}^-} \left(\dfrac{1}{\cos x} - \dfrac{\sin x}{\cos x} \right)$

$$= \lim\limits_{x \to \frac{\pi}{2}^-} \dfrac{1 - \sin x}{\cos x} \left(\text{form } \dfrac{0}{0} \right) = \lim\limits_{x \to \frac{\pi}{2}^-} \dfrac{-\cos x}{-\sin x} = 0.$$

Forms 0^0, 1^∞, ∞^0:

Finally, for limits of the type $\lim\limits_{x \to a} (f(x))^{g(x)}$, indeterminate forms of type 0^0, 1^∞, and ∞^0 are possible. In these cases,

1. Let $y = (f(x))^{g(x)}$.
2. Then $\ln y = g(x) \ln f(x)$.
3. If $\lim\limits_{x \to a} g(x) \ln f(x)$ (form $0 \cdot \infty$) exists and equals L,

   $$\lim\limits_{x \to a} (f(x))^{g(x)} = e^L.$$

Section 3.9: Indeterminate Forms and L'Hospital's Rule

1) True, False:

   All limits having one of the indeterminate forms are solved by reducing (if necessary) to limits of the form $\frac{0}{0}$ or $\frac{\infty}{\infty}$.

   <u>True.</u> L'Hospital's Rule only applies in $\frac{0}{0}$ and $\frac{\infty}{\infty}$ cases.

2) Evaluate each:

   a) $\displaystyle\lim_{x \to 3} \frac{x^2 - 7x + 12}{x^2 - 9}$

   $\displaystyle\lim_{x \to 3} \frac{x^2 - 7x + 12}{x^2 - 9}$ (form $\frac{0}{0}$)

   $= \displaystyle\lim_{x \to 3} \frac{2x - 7}{2x} = \frac{-1}{6}$.

   b) $\displaystyle\lim_{x \to 0} \frac{\sin x}{x}$

   $\displaystyle\lim_{x \to 0} \frac{\sin x}{x}$ (a familiar limit)

   $= \displaystyle\lim_{x \to 0} \frac{\cos x}{1} = \frac{1}{1} = 1$.

   c) $\displaystyle\lim_{x \to \infty} \frac{e^x}{3x}$

   $\displaystyle\lim_{x \to \infty} \frac{e^x}{3x}$ (form $\frac{\infty}{\infty}$)

   $= \displaystyle\lim_{x \to \infty} \frac{e^x}{3} = \infty$.

d) $\lim\limits_{x \to 0^+} x \cot x$

This has form $0 \cdot \infty$. Rewrite $x \cot x$ as

$$x \cdot \frac{\cos x}{\sin x} = \frac{x}{\sin x} \cdot \cos x.$$

$$\lim_{x \to 0^+} x \cot x = \lim_{x \to 0^+} \frac{x}{\sin x} \cdot \cos x$$

$$= \left(\lim_{x \to 0^+} \frac{x}{\sin x} \right) \cdot \left(\lim_{x \to 0^+} \cos x \right) \left(\text{form } \frac{0}{0} \right)$$

$$= \lim_{x \to 0^+} \frac{1}{\cos x} \cdot \lim_{x \to 0^+} \cos x$$

$$= \lim_{x \to 0^+} 1 = 1.$$

e) $\lim\limits_{x \to \infty} \dfrac{1}{x^2 - 1} - \dfrac{1}{x - 1}$

This limit is $0 - 0 = 0$. (Not an indeterminate form.)

f) $\lim\limits_{x \to 0^+} \left(\dfrac{1}{x} - \dfrac{1}{e^x - 1} \right)$

$$\lim_{x \to 0^+} \left(\frac{1}{x} - \frac{1}{e^x - 1} \right) \ (\text{form } \infty - \infty)$$

$$= \lim_{x \to 0^+} \frac{e^x - 1 - x}{x(e^x - 1)} \ (\text{form } \frac{0}{0})$$

$$= \lim_{x \to 0^+} \frac{e^x - 1}{xe^x + e^x - 1} \ \left(\text{form } \frac{0}{0} \text{ again} \right)$$

$$= \lim_{x \to 0^+} \frac{e^x}{xe^x + 2e^x} = \frac{1}{0 + 2} = \frac{1}{2}.$$

g) $\displaystyle\lim_{x \to 0^+} \left(\frac{1}{\sin^2 x} - \frac{\cot x}{x} \right)$

$$\lim_{x \to 0^+} \left(\frac{1}{\sin^2 x} - \frac{\cot x}{x} \right)$$

$$= \lim_{x \to 0^+} \frac{x - \cos x \sin x}{x \sin^2 x} \quad (\text{form } \frac{0}{0})$$

$$= \lim_{x \to 0^+} \frac{1 - \cos^2 x + \sin^2 x}{\sin^2 x + 2x \sin x \cos x}$$

$$= \lim_{x \to 0^+} \frac{2 \sin^2 x}{\sin x (\sin x + 2x \cos x)}$$

$$= \lim_{x \to 0^+} \frac{2 \sin x}{\sin x + 2x \cos x} \quad \left(\text{form } \frac{0}{0} \right)$$

$$= \lim_{x \to 0^+} \frac{2 \cos x}{3 \cos x - 2x \sin x} = \frac{2}{3}.$$

h) $\displaystyle\lim_{x \to 0^+} (\csc x)^x$

This has form ∞^0. Let $y = (\csc x)^x$.

$\ln y = x \ln (\csc x)$.

$$\lim_{x \to 0^+} x (\ln \csc x) \quad (\text{form } 0 \cdot \infty)$$

$$= \lim_{x \to 0^+} \frac{\ln \csc x}{x^{-1}} \quad (\text{form } \frac{\infty}{\infty})$$

$$= \lim_{x \to 0^+} \frac{\dfrac{1}{\csc x} \cdot (- \csc x \cot x)}{(-1)x^{-2}}$$

$$= \lim_{x \to 0^+} x^2 \cot x \quad (\text{form } 0 \cdot \infty)$$

$$= \lim_{x \to 0^+} \frac{x^2}{\tan x} \quad (\text{form } \frac{0}{0})$$

$$= \lim_{x \to 0^+} \frac{2x}{\sec^2 x} = \frac{2(0)}{1} = 0.$$

Thus $\displaystyle\lim_{x \to 0^+} (\csc x)^x = e^0 = 1.$

4

The Mean Value Theorem & Curve Sketching

"GRANTZ IS CHARTING HIS LIFE BASED ON
GENETIC VS. ENVIRONMENTAL FACTORS."

Maximum and Minimum Values

Concepts to Master

A. Absolute maxima, minima, extrema; Relative (local) maxima, extrema;
Extreme Value Theorem

B. Critical Numbers; Fermat's Theorem about local extrema;
Absolute extrema of a continuous function on a closed interval

Summary and Focus Questions

A. Let f be a function with domain D. f has an <u>absolute maximum</u> at c (and $f(c)$ is the <u>maximum value</u>) means $f(x) \leq f(c)$ for all $x \in D$. Thus a highest point on the graph of f occurs at $(c, f(c))$.

f has an <u>absolute minimum</u> at c means $f(x) \geq f(c)$ for all $x \in D$, so $(c, f(c))$ is a lowest point on the graph.

The <u>extreme values</u> of f are the maximum of f (if there is one) and the minimum of f (if there is one). Extreme values are important to calculate (viz., maximum profit, minimum force, etc.) The following three graphs from left to right have two, one and no extreme values, respectively.

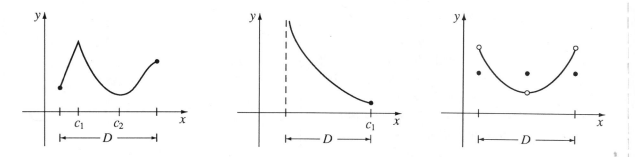

The <u>Extreme Value Theorem</u> says that the extreme values must always exist for a continuous function whose domain is a closed interval.

If there is some open interval I containing c such that $f(c) \geq f(x)$ for all $x \in I$, then f has a <u>local (or relative) maximum</u> at c. In case $f(c) \leq f(x)$ for all $x \in I$, we say f has a <u>local (or relative) minimum</u> at c.

The following graph is a function with domain $[a, b]$ which has local maxima at $x = c, p, r, t$ and local minima at $x = a, d, q, s, b$.

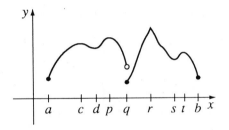

1) If $f(x) \geq f(c)$ for all x in the domain of the function f, then f has an _____ at c.

absolute maximum

2) True or False:

A function may have more than one absolute minimum value.

<u>False.</u>

3) True or False:

The absolute minimum value of a function may occur at more than one point.

True. For example, $f(x) = 9 - x^2$ with domain $[-2, 2]$, has a minimum value of 5 which occurs at both $x = 2$ and $x = -2$.

4) Does the Extreme Value Theorem guarantee that the extreme values exist for the the function graphed?

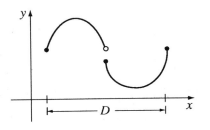

No. The function is not continuous on D. In this case the extreme values still exist.

5) True or False:
   If f has an absolute minimum at $x = c$ then f has a local minimum at $x = c$.

True.

6) How many local maxima and minima does the following function with domain $[a, b]$ have?

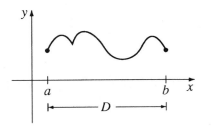

3 local maxima, and 4 local minima.

B. A <u>critical number</u> c is a number in the domain of f for which either $f'(c) = 0$ or $f'(c)$ does not exist.

To determine where $f'(x) = 0$ is a matter of solving the equation $f'(x) = 0$. The method of solution depends greatly on the nature of the function. Factoring will sometimes work as in this example.

<u>Example:</u>

Solve $f'(x) = 0$ where $f(x) = 3x^4 + 20x^3 - 36x^2 + 7$.

$f'(x) = 12x^3 + 60x^2 - 72x = 12x(x - 1)(x + 6)$. So $f'(x) = 0$ at $x = 0$, $x = 1$, $x = -6$.

To determine where $f'(x)$ does not exist is often (but not always!) a matter of finding points where a denominator is 0 or where an even root of a negative number occurs.

<u>Examples:</u>

Find all points in the domain of f at which $f'(x)$ does not exist.

a) $f(x) = 9\sqrt{x} + x^{3/2}$.

Here the domain is $[0, \infty]$, $f'(x) = \dfrac{9}{2\sqrt{x}} + \dfrac{3\sqrt{x}}{2}$ and $f'(x)$ does not exist for $x = 0$.

b) $f(x) = \sqrt{4 - x}$.

The domain of f is $(-\infty, 4]$ and $f'(x) = \dfrac{-1}{2\sqrt{4 - x}}$ does not exist at $x = 4$.

<u>Fermat's Theorem</u> states that if f has a local maximum or minimum at c and $f'(c)$ exists, then $f'(c) = 0$. In other words, if $f(c)$ is a local extremum for f, then c is a critical number of f. The converse is false.

The extreme values of a continuous function f on a closed interval $[a, b]$ always occur either at a, at b, or at a critical number of f in $[a, b]$.

7) Suppose $f'(c) = 0$. Is c a local maximum or minimum?

Not necessarily; for example,
$f(x) = x^3$ with $c = 0$ has $f'(0) = 0$ but $c = 0$
is not a local extremum of $f(x) = x^3$.

8) Find the extreme values of
$f(x) = \sqrt{10x - x^2}$ on $[2, 10]$.

Since f is continuous on $[2, 10]$, the extreme values occur at 2, 10, or some critical number between 2 and 10.
$$f(x) = (10x - x^2)^{1/2}$$
$$f'(x) = \frac{1}{2}(10x - x^2)^{-1/2}(10 - 2x)$$
$$= \frac{5 - x}{\sqrt{10x - x^2}}.$$

On $[2, 10]$, $f'(x) = 0$ at $x = 5$.
$f'(x)$ does not exist at $x = 10$.
Thus 5 is the only critical number between 2 and 10.
Computing functional values:

x	5	2	10
$f(x)$	5	4	0

The absolute minimum is 0 at $x = 10$ and the absolute maximum is 5 at $x = 5$.

The Mean Value Theorem

Concepts to Master

Rolle's Theorem; Mean Value Theorem

Summary and Focus Questions

The Mean Value Theorem is very important because it contains an equation that relates function values and derivative values. Rolle's Theorem is a special case:

Rolle's Theorem:

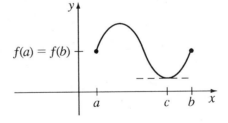

> If the function f
> 1) is continuous on $[a, b]$
> 2) is differentiable on (a, b)
> 3) has $f(a) = f(b)$
> then there is at least one number
> $c \in (a, b)$ such that $f'(c) = 0$.

Mean Value Theorem:

> If the function f
> 1) is continuous on $[a, b]$
> 2) is differentiable on (a, b)
> then there is at least one $c \in (a, b)$
>
> such that $f'(c) = \dfrac{f(b) - f(a)}{b - a}$.

Since $\dfrac{f(b) - f(a)}{b - a}$ is the slope of the line joining the endpoints of the graph, the Mean

Value Theorem simply says there is at least one point in between the endpoints where the tangent line has the same slope as that of the line joining the endpoints.

If $f'(x) = g'(x)$ for all $x \in (a, b)$, then $f(x) = g(x) + c$ where c is a constant.

1) For $f(x) = 1 - x^2$ on $[-2, 1]$, do the hypotheses and conclusion of Rolle's Theorem hold?

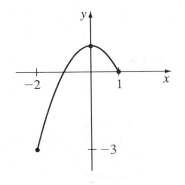

f is continuous on $[-2, 1]$ and differentiable on $(-2, 1)$ but since $f(-2) = -3 \neq 0 = f(1)$ one hypothesis fails. The conclusion holds because $0 \in (-2, 1)$ and $f'(0) = 0$.

2) Do the hypotheses and conclusion of the Mean Value Theorem hold for $f(x) = 6x + x^2$ on $[1, 3]$?

The hypotheses hold because f is a polynomial and therefore continuous and differentiable everywhere. Since the hypotheses are true, the conclusion is true as well.

$$\frac{f(3) - f(1)}{3 - 1} = \frac{27 - 7}{3 - 1} = 10$$

$f'(x) = 6 + 2x$ which equals 10 for $c = 2$ in $(1, 3)$.

Section 4.2: The Mean Value Theorem

3) Joe is driving on a highway which has a speed limit of 55 mi/hr. At 2 P.M. he is at milepost 110 and at 5 P.M. he is at milepost 290. Could Joe be proved guilty of speeding?

Yes. Let $f(t)$ be his position at time t. $f(2) = 110$ and $f(5) = 290$. The mean value on $[2, 5]$ is

$$\frac{f(5) - f(2)}{5 - 2} = \frac{290 - 110}{5 - 2} = 60.$$

By the Mean Value Theorem there is a time $c \in (2, 5)$ such that $f'(c) = 60$. So at least once (at time c) he was going 60 mph.

4) Find all numbers c that satisfy the conclusion of the Mean Value Theorem for $f(x) = x^3 - 3x^2 + x$ on $[0, 3]$.

The mean value on $[0, 3]$ is $\dfrac{f(3) - f(0)}{3 - 0}$

$$= \frac{3 - 0}{3 - 0} = 1.$$

$f'(x) = 3x^2 - 6x + 1.$
$3x^2 - 6x + 1 = 1$
$3x^2 - 6x = 0$
$3x(x - 2) = 0.$
Thus $f'(x) = 1$ at $x = 0, x = 2$.
However, $0 \notin (0, 3)$ so $c = 2$ is the only point satisfying the Mean Value Theorem.

5) True, False:
If $f'(x) = g'(x)$, then $f(x) = g(x)$.

False. For example, $f(x) = x^2$ and $g(x) = x^2 + 6$ have the same derivative, $2x$.

Monotonic Functions and the First Derivative Test

Concepts to Master

A. Increasing, decreasing, monotonic functions; Relationship of the derivative to increasing and decreasing

B. The First Derivative Test

Summary and Focus Questions

A. A function f is <u>increasing</u> on an interval I if for all $x_1, x_2, \in I$, $x_1 < x_2$ implies $f(x_1) < f(x_2)$. <u>Decreasing</u> is defined as $x_1 < x_2$ implies $f(x_1) > f(x_2)$ for all $x_1, x_2, \in I$. f is <u>monotonic</u> on I if f is either increasing on I or decreasing on I.

If f is continuous on $[a, b]$ and differentiable on (a, b):
   a) $f'(x) > 0$ for all $x \in (a, b)$ implies f is increasing on $[a, b]$.
   b) $f'(x) < 0$ for all $x \in (a, b)$ implies f is decreasing on $[a, b]$.

1) Is $f(x) = x^2 + 6x$ increasing on $[-1, 2]$?

<u>Yes.</u> $f'(x) = 2x + 6$. $2x + 6 > 0$, when $2x > -6$ or $x > -3$.
Thus for all $x \in [-1, 2]$, $f'(x) > 0$.

2) Where is $f(x) = x^3 - 3x^2$ increasing and where is it decreasing?

$f'(x) = 3x^2 - 6x = 3x(x - 2) = 0$ at $x = 0, x = 2$.

Section 4.3: Monotonic Functions and the First Derivative Test

	$3x$	$x - 2$	$3x(x - 2)$
$x < 0$	$-$	$-$	$+$
$0 < x < 2$	$+$	$-$	$-$
$2 < x$	$+$	$+$	$+$

f is increasing on $(-\infty, 0)$ and $(2, \infty)$;
f is decreasing on $(0, 2)$.

3) Sketch a graph of a differentiable function having all these properties:

$f(0) = 1, f(2) = 3, f(5) = 0,$
$f'(x) > 0$ for $x \in (0, 2)$ and $x \in (5, \infty),$
$f'(x) < 0$ for $x \in (-\infty, 0)$ and $x \in (2, 5).$

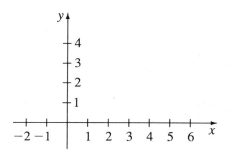
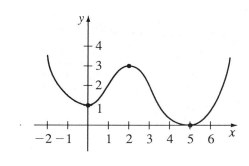

B. <u>The First Derivative Test</u>:

Let f be continuous on $[a, b]$ and differentiable on (a, b). Let c be a critical number for f in (a, b).

a) If $f'(x)$ is positive in (a, c) and negative in (c, b), then f has a local maximum at c.

b) If $f'(x)$ is negative in (a, c) and positive in (c, b), f has a local minimum at c.

c) If $f'(x)$ has the same sign (either positive or negative) in both (a, c) and (c, b), $f(c)$ is not a local extrema.

4) Suppose 8 is a critical number for a
   function f with $f'(x) > 0$ for $x \in (8, 9)$
   and $f'(x) < 0$ for $x \in (7, 8)$. Then 8 is a
   a) local maximum
   b) local minimum
   c) not a local extremum.

b) By the First Derivative Test.

5) What can you conclude about the local
   extrema of a differential function f from
   the following?

Interval	Sign of $f'(x)$
$(-\infty, 1)$	positive
$(1, 3)$	negative
$(3, 4)$	negative
$(4, 6)$	positive
$(6, 8)$	positive
$(8, \infty)$	negative

f has a local maximum at $x = 1$ and $x = 8$;
f has a local minimum at $x = 4$; there are
no local extrema at $x = 3$ or $x = 6$.

6) Is $x = 5$ a local minimum for
   $f(x) = 10x - x^2 + 1$?

No. $f'(x) = 10 - 2x > 0$ for all $x < 5$
while $f'(x) < 0$ for all $x > 5$. f has a local
maximum at $x = 5$.

Concavity and Points of Inflection

Concepts to Master

A. Concave upward and downward; Test for concavity
B. Point of inflection; The Second Derivative Test

Summary and Focus Questions

A. A function is <u>concave upward on an interval</u> I if f lies above all tangent lines to f in I.

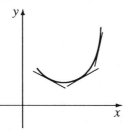

f is <u>concave downward on</u> I if the graph of f is below all tangent lines.

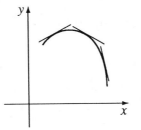

A test for concavity involves the second derivative:

If f is continuous on $[a, b]$ and twice differentiable on (a, b) (meaning $f''(x)$ exists for $x \in (a, b)$) then
    a) if $f''(x) > 0$ for all $x \in (a, b)$, f is concave *upward* on $[a, b]$.
    b) if $f''(x) < 0$ for all $x \in (a, b)$, f is concave *downward* on $[a, b]$.

– 135 –

1) Use the graph below to answer true or
   false to each.

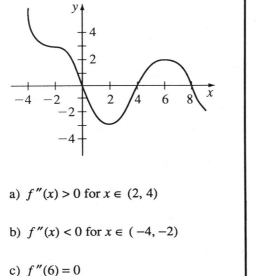

a) $f''(x) > 0$ for $x \in (2, 4)$

True.

b) $f''(x) < 0$ for $x \in (-4, -2)$

False.

c) $f''(6) = 0$

False. $f''(6)$ is negative.

d) $f''(2) > 0$

True.

e) f is concave upward on $(0, 2)$

True.

B. A <u>point of inflection</u> for f is a point on the graph of f where concavity changes
from concave downward to concave upward or from concave upward to concave
downward.

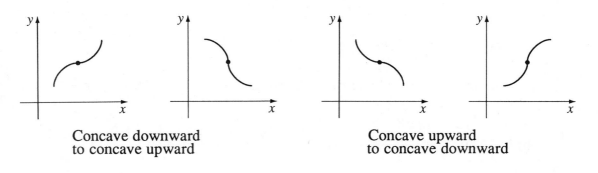

Concave downward
to concave upward

Concave upward
to concave downward

The <u>Second Derivative Test</u> is:

If f'' is continuous on (a, b) containing c and $f'(c) = 0$, then
　　a) if $f''(c) < 0$, f has a local maximum at c.
　　b) if $f''(c) > 0$, f has a local minimum at c.

If it should happen that both $f'(c) = 0$ and $f''(c) = 0$ then no conclusion can be made about an extremum at c.

2) What are the points of inflection for the function in question 1)?

$x = -2, 0, 4, 8$.

3) Find the points of inflection for
$f(x) = 8x^3 - x^4$.

$f'(x) = 24x^2 - 4x^3$,
$f''(x) = 48x - 12x^2 = 12x(4 - x) = 0$
at $x = 0$, $x = 4$. Since f is a polynomial
the only points of inflection are at 0 and 4.

4) Find all local extrema of
　　a) $f(x) = x^3 - 6x^2 - 36x$.

$f'(x) = 3x^2 + 12x - 36 = 3(x - 2)(x + 6)$.
The critical points are at $x = 2, -6$.
$f''(x) = 6x + 12$.
At $x = 2$, $f''(2) = 24 > 0$ so f has a
local minimum at 2.
At $x = -6$, $f''(-6) = -24 < 0$, so f
has a local maximum at -6.

　　b) $f(x) = x^2 + 2 \cos x$.

$f'(x) = 2x - 2 \sin x = 0$ when $x = \sin x$. The
only solution to this equation is $x = 0$.
$f''(x) = 2 - 2 \cos x$ but $f''(0) = 0$ so
the Second Derivative Test fails to give
additional information. However
$f''(x) \geq 0$, for all x, means f is concave
upward and thus f has a local minimum at 0.

Curve Sketching

Concepts to Master

A. Curve sketching using information obtained through calculus concepts
B. Slant asymptotes

Summary and Focus Questions

A. Questions to answer in sketching a graph of $y = f(x)$ include:

A. Domain of f — For what x is $f(x)$ defined?

B. x intercepts — What are the solution(s) (if any) to $f(x) = 0$?

    y intercepts — What value (if any) is $f(0)$?

C. Symmetry about y-axis — Is $f(-x) = f(x)$?

    Symmetry about origin — Is $f(-x) = -f(x)$?

    Periodic — Is there a number p such that $f(x + p) = f(x)$ for all x in the domain?

D. Horizontal asymptote(s) — Does $\lim\limits_{x \to \infty} f(x)$ or $\lim\limits_{x \to -\infty} f(x)$ exist?

    Vertical asymptote(s) — For what a is $\lim\limits_{x \to a^+} f(x) = \infty$ or $-\infty$?

    For what a is $\lim\limits_{x \to a^-} f(x) = \infty$ or $-\infty$?

E. Increasing   On what intervals is $f'(x) \geq 0$?

    Decreasing   On what intervals is $f'(x) \leq 0$?

F. Critical numbers   Where does $f'(x) = 0$ or not exist?

    Local extrema   Where are the local maxima or minima (if any)?

                      (Use the First and Second Derivative Tests)

G. Concave upward   On what intervals is $f''(x) \geq 0$?

    Concave downward   On what intervals is $f''(x) \leq 0$?

    Inflection points   Where does f change concavity?

                      Where is $f''(x) = 0$?

1) Sketch a graph of $y = f(x)$ that has all these properties:

    a) domain $= (-\infty, -1) \cup (1, \infty)$

    b) $f(2) = 0$

    c) $f(x) = -f(-x)$

    d) $\lim\limits_{x \to \infty} f(x) = 4$ and $\lim\limits_{x \to 1^+} f(x) = -\infty$

    e) $f'(3) = 0,\ f'(5) = 0$

    f) $f'(x) \geq 0$ on $(1, 3) \cup (5, \infty)$
      $f'(x) \leq 0$ on $(3, 5)$

    g) local maximum at 3
      local minimum at 5

    h) $f''(x) \geq 0$ on $(4, 6)$
      $f''(x) \leq 0$ on $(1, 4) \cup (6, \infty)$

    i) $f''(4) = 0, f''(6) = 0$

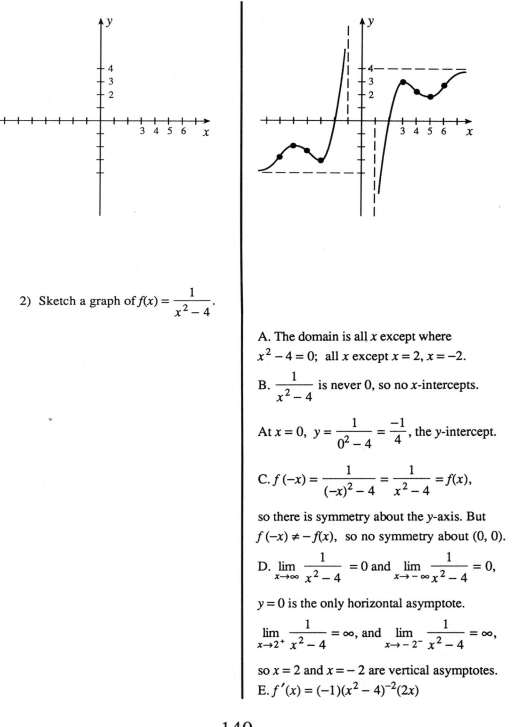

2) Sketch a graph of $f(x) = \dfrac{1}{x^2 - 4}$.

A. The domain is all x except where $x^2 - 4 = 0$; all x except $x = 2$, $x = -2$.

B. $\dfrac{1}{x^2 - 4}$ is never 0, so no x-intercepts.

At $x = 0$, $y = \dfrac{1}{0^2 - 4} = \dfrac{-1}{4}$, the y-intercept.

C. $f(-x) = \dfrac{1}{(-x)^2 - 4} = \dfrac{1}{x^2 - 4} = f(x)$,

so there is symmetry about the y-axis. But $f(-x) \neq -f(x)$, so no symmetry about $(0, 0)$.

D. $\displaystyle\lim_{x \to \infty} \dfrac{1}{x^2 - 4} = 0$ and $\displaystyle\lim_{x \to -\infty} \dfrac{1}{x^2 - 4} = 0$,

$y = 0$ is the only horizontal asymptote.

$\displaystyle\lim_{x \to 2^+} \dfrac{1}{x^2 - 4} = \infty$, and $\displaystyle\lim_{x \to -2^-} \dfrac{1}{x^2 - 4} = \infty$,

so $x = 2$ and $x = -2$ are vertical asymptotes.

E. $f'(x) = (-1)(x^2 - 4)^{-2}(2x)$

$$= -2x(x^2 - 4)^{-2} = \frac{-2x}{(x^2 - 4)^2}.$$

For $x > 0$, $x \neq 2$, $f'(x) < 0$ so f is decreasing on $(0, 2)$ and $(2, \infty)$.
For $x < 0$, $x \neq -2$, $f'(x) > 0$ so f is increasing on $(-\infty, -2)$ and $(2, 0)$.

F. $f'(x) = \dfrac{-2x}{(x^2 - 4)^2} = 0$ at $x = 0$.

$f''(x) =$
$$(-2x)[-2(x^2 - 4)^{-3} \cdot 2x] + (x^2 - 4)^{-2}(-2)$$
$$= (x^2 - 4)^{-3}[8x^2 - 2(x^2 - 4)] = \frac{6x^2 + 4}{(x^2 - 4)^3}.$$

$f''(0) = \dfrac{4}{(-4)^3} = -\dfrac{1}{16}$. Thus f has a

local maximum at $x = 0$.

G. Since $f''(x) = \dfrac{6x^2 + 4}{(x^2 - 4)^3}$, the sign of

$f''(x)$ is determined by $(x^2 - 4)^3$ and thus by $x^2 - 4$. $x^2 - 4 > 0$ for $x > 2$ and $x < -2$ and $x^2 - 4 < 0$ for $-2 < x < 2$. f is concave upward on $(-\infty, -2)$ and $(2, \infty)$ and concave downward on $(-2, 2)$.
H. Finally, a sketch:

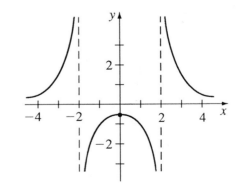

3) Sketch a graph of $f(x) = x - \ln x$.

A. The domain is $(0, \infty)$.

B. $x - \ln x$ is never 0, so no x-intercept.

C. Since $x > 0$ there is no symmetry.

D. $\lim\limits_{x \to 0^+} x - \ln x = \infty$

$\lim\limits_{x \to \infty} x - \ln x = \infty$.

E. $f'(x) = 1 - \dfrac{1}{x}$

For $0 < x < 1, f'(x) < 0$ and f is decreasing.

For $x > 1, f'(x) > 0$ and f is increasing.

F. $f'(x) = 1 - \dfrac{1}{x} = 0$ at $x = 1$.

$f''(x) = x^{-2}$, so $f''(1) = 1 > 0$.

Thus f has a local minimum at $x = 1$.

G. Since $f''(x) > 0$ the graph is always concave upward.

H. Here is a sketch:

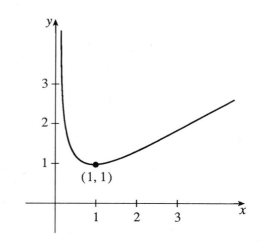

B. $y = mx + b$ is a <u>slant asymptote</u> of $y = f(x)$ means that $\lim\limits_{x \to \infty} [f(x) - (mx + b)] = 0$

or $\lim\limits_{x \to -\infty} [f(x) - (mx + b)] = 0$.

A slant asymptote is a nonvertical, nonhorizontal line that the curve approaches for large (positive or negative) values of x.

3) Find a slant asymptote for
$$f(x) = \frac{x^2 + x + 1}{x}.$$

$f(x) = \dfrac{x^2 + x + 1}{x} = x + 1 + \dfrac{1}{x}$. The line

$y = x + 1$ is a slant asymptote.

4) Can a polynomial of degree greater than one have a slant asymptote?

<u>No.</u>

Applied Maximum and Minimum Problems

Concepts to Master

Solutions to applied extremum problems

Summary and Focus Questions

Many problems can be characterized as "find the maximum (or minimum) value of some quantity Q subject to some given conditions." A procedure to solve some such problems is:

(1) Introduce notation identifying the quantity to be maximized (or minimized), Q, and other variable(s).
(2) Write Q as a function of the other variables. Here a diagram may help.
(3) Rewrite, if necessary, Q as a function of just one of the variables, using given relationships and determine the domain D of Q.
(4) Find the absolute extrema of Q on domain D.

Steps 2 and 3 are often the most difficult because you may need to recall facts from geometry, trigonometry, etc.

1) A gardener wishes to enclose a rectangular area with 300 feet of fence and fence it down the middle to divide it into two equal subareas. What is the largest rectangular area that may be enclosed?

Section 4.6: Applied Maximum and Minimum Problems

(1) Let A be the area of the enclosed rectangle.

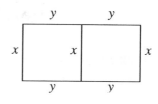

Label one side as x and the half side of the other as y (since that side is bisected).

(2) $A = x(2y) = 2xy$.

(3) The total amount of fence is

$3x + 4y = 300$. Thus $y = \dfrac{(300 - 3x)}{4}$.

Therefore $A = 2xy = 2x\dfrac{(300 - 3x)}{4}$.

$= 150x - \dfrac{3}{2}x^2$. Since $3x + 4y = 300$ and both $x \geq 0$ and $y \geq 0$, $x \in [0, 100]$.

(4) Maximize $A = 150x - \dfrac{3}{2}x^2$ on $[0, 100]$. Clearly 0 and 100 produce minimum values, so A is maximum when $A' = 0$.
$A' = 150 - 3x = 0$ at $x = 50$.

Then $y = \dfrac{300 - 3(50)}{4} = 37.5$.

Thus the maximum value of A is
$2xy = 2(50)(37.5) = 3750$ ft^2.

Applications to Economics

Concepts to Master

A. Determining minimum average cost
B. Determining maximum profit

Summary and Focus Questions

A. If $C(x)$ represents the (total) cost function to produce x items, the <u>average cost</u> function is:

$$c(x) = \frac{C(x)}{x}$$

<u>Marginal cost</u> is the derivative $C'(x)$ of total cost.

Average cost $c(x)$ is minimum when $c(x) = C'(x)$.

1) Find the minimum average cost for a company which produces x items at a total of $3x^2 + 10x + 1875$ dollars.

$C(x) = 3x^2 + 10x + 1875$

$C'(x) = 6x + 10$

$c(x) = \dfrac{3x^2 + 10x + 1875}{x}$

$\qquad = 3x + 10 + \dfrac{1875}{x}.$

Average cost is minimum when $c(x) = C'(x)$:

$3x + 10 + \dfrac{1875}{x} = 6x + 10$

$$\frac{1875}{x} = 3x$$

$$x^2 = 625$$

$x = 25$ ($x = -25$ is not a meaningful answer).

Thus the minimum average cost is

$$3(25) + 10 + \frac{1875}{25} = \$160 \text{ per unit.}$$

B. The <u>price (or demand) function</u> $d(x)$ is the price per item a firm must charge to sell x items. The (total) <u>revenue function</u> $R(x) = x\,d(x)$ is the revenue obtained by selling x items.

<u>Marginal revenue</u> is the derivative $R'(x)$ of total revenue.

The <u>profit function</u> $P(x) = R(x) - C(x)$ is a maximum when marginal revenue equals marginal cost.

2) Suppose a laundry determines that to attract x customers per day its price of service must be $3.20 - .02x$ dollars. If their total cost to serve x customers is $.05x^2 + 1.10 + 120$ dollars, how many customers should be served to maximize their profit?

(1) The demand is $p(x) = 3.20 + .02x$
so the revenue is $R(x) = p \cdot x = 3.20x + .02x^2$.
Thus marginal revenue is $R'(x) = 3.20 + .04x$.
(2) Total cost is $C(x) = .05x^2 + 1.10x + 120$,
so marginal cost is $C'(x) = .10x + 1.10$.
(3) Equate $R'(x)$ and $C'(x)$:
$3.20 + .04x = .10x + 1.10$
$2.10 = .06x$
$x = 35$ customers.

Antiderivatives

Concepts to Master

General and particular antiderivatives of a function

Summary and Focus Questions

$F(x)$ is an <u>antiderivative</u> of the function $f(x)$ means $F'(x) = f(x)$.
If F is an antiderivative of f then <u>all</u> other antiderivatives of f have the form $F(x) + C$,
where C is a constant.

Function	Form of All Antiderivatives
x^n (except $n = -1$)	$\dfrac{x^{n+1}}{n+1} + C$
$\sin x$	$-\cos x + C$
$\cos x$	$\sin x + C$
$\sec^2 x$	$\tan x + C$
$\sec x \tan x$	$\sec x + C$
$1/x$	$\ln|x| + C$
e^x	$e^x + C$
$\dfrac{1}{\sqrt{1-x^2}}$	$\sin^{-1} x$
$\dfrac{1}{1+x^2}$	$\tan^{-1} x$

In addition, the antiderivative of $f \pm g$ is the antiderivative of f plus or minus the antiderivative of g, and the antiderivative of $c\,f(x)$ is c times the antiderivative of $f(x)$ (c, any constant).

The general solution to the differential equation $\frac{dy}{dx} = f(x)$ is all antiderivatives of f.

To find a particular solution, first find the general solution, then substitute the given values to determine the constant C.

For example, if $f'(x) = 6x$ and $f(1) = 7$, then $f(x) = 2x^3 + C$ where C is a constant. $f(1) = 2(1)^3 + C = 2 + C$. From $2 + C = 7$, $C = 5$. The particular solution is $f(x) = 2x^3 + 5$.

1) If $g(x)$ is an antiderivative of $h(x)$, then

_____ $'(x) =$ _____ (x).

$g'(x) = h(x)$.

2) Find all antiderivatives of

a) $f(x) = x^7$

$\dfrac{x^8}{8} + C$

b) $f(x) = \cos x - \sec^2 x$

$\sin x - \tan x + C$

c) $f(x) = x + x^{-2}$

$\dfrac{x^2}{2} - \dfrac{1}{x} + C$

d) $f(x) = \dfrac{2}{x}$

$2 \ln |x| + C$

e) $f(x) = \dfrac{-1}{1 + x^2}$

$-\tan^{-1} x + C$

3) Find $f(x)$ where $f(2) = 3$ and $f'(x) = 4x + 5$.

$f(x) = 2x^2 + 5x + C$
$f(2) = 2(2)^2 + 5(2) + C = 18 + C$
$18 + C = 3$ so $C = -15$
$f(x) = 2x^2 + 5x - 15$.

5

Integrals

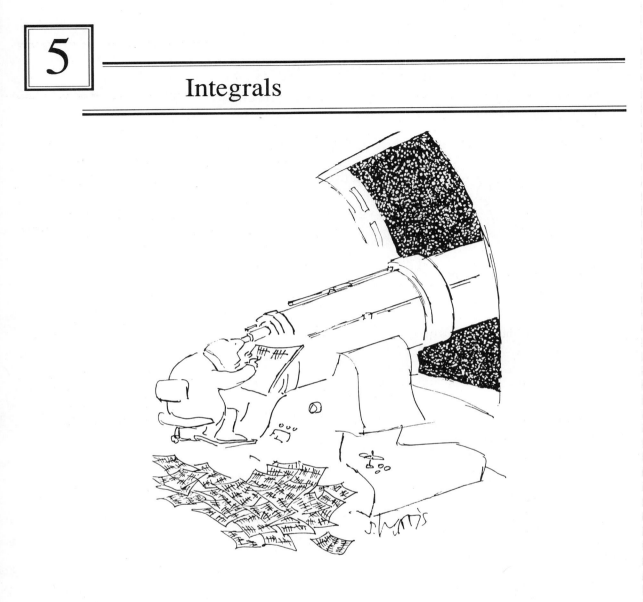

Sigma Notation

Concepts to Master

Evaluation of sums using sigma notation

Summary and Focus Questions

For real numbers a_m, $a_{m+1}, \ldots a_n$,

$$\sum_{i=m}^{n} a_i = a_m + a_{m+1} + \ldots + a_n.$$

This is a compact notation for the sum of several numbers.

$\sum_{i=m}^{n} f(i)$ is the number found by determining the numbers $f(m), f(m+1), \ldots, f(n)$

and then adding them up, as in:

$$\sum_{i=2}^{5} i^2 = (2)^2 + (3)^2 + (4)^2 + (5)^2 = 4 + 9 + 16 + 25 = 54.$$

Some rules for manipulating sigma notation sums are:

1. $\displaystyle\sum_{i=m}^{n} c\, a_i = c\left(\sum_{i=m}^{n} a_i\right)$, where c is a constant.

2. $\displaystyle\sum_{i=m}^{n} (a_i \pm b_i) = \sum_{i=m}^{n} a_i \pm \sum_{i=m}^{n} b_i.$

Some summations may be rewritten as a function of their upper limit thus replacing the summation notation with a simple expression.

$$\sum_{i=1}^{n} 1 = n \qquad \sum_{i=1}^{n} i = \frac{n(n+1)}{2} \qquad \sum_{i=1}^{n} i^2 = \frac{n(n+1)(2n+1)}{6}$$

$$\sum_{i=1}^{n} i^3 = \left(\frac{n(n+1)}{2}\right)^3 \qquad \sum_{i=1}^{n} i^4 = \frac{n(n+1)(2n+1)(3n^2+3n-1)}{30}$$

These may be combined to simplify other sums as in:

$$\sum_{i=1}^{n} (i^2 + 4i) = \sum_{i=1}^{n} i^2 + \sum_{i=1}^{n} 4i = \sum_{i=1}^{n} i^2 + 4\sum_{i=1}^{n} i$$

$$= \frac{n(n+1)(2n+1)}{6} + 4\frac{n(n+1)}{2} = \frac{n(n+1)(2n+13)}{6}.$$

1) Write the following in sigma notation.
   a) $2^2 + 3^2 + 4^2 + 5^2 + 6^2$.

$$\sum_{i=2}^{6} i^2$$

   b) $\frac{1}{2} - \frac{1}{4} + \frac{1}{8} - \frac{1}{16} + \frac{1}{32}$.

$$\sum_{i=1}^{5} \frac{(-1)^{i+1}}{2^i} \quad \left(\text{or} - \sum_{i=1}^{5} \left(-\frac{1}{2}\right)^i\right)$$

2) What number is $\displaystyle\sum_{i=1}^{4} 3i$?

From the definition, this is

$3(1)+3(2)+3(3)+3(4) = 3 + 6 + 9 + 12 = 30.$

It may also be evaluated as

$$\sum_{i=1}^{4} 3i = 3 \sum_{i=1}^{4} i = 3\left[\frac{4(4+1)}{2}\right] = 30.$$

3) Write the following as an expression in terms of n :

$$\sum_{i=1}^{n} (12i^2 - 4ni)$$

$$\sum_{i=1}^{n} (12i^2 - 4ni) = \sum_{i=1}^{n} 12i^2 - \sum_{i=1}^{n} 4ni$$

$$= 12 \sum_{i=1}^{n} i^2 - 4n \sum_{i=1}^{n} i$$

$$= 12 \frac{n(n+1)(2n+1)}{6} - (4n)\frac{n(n+1)}{2}$$

$$= n(n+1)(2n+2).$$

4) Evaluate $\sum_{i=1}^{10} (f(i) - f(i-1))$, where

$$f(i) = 2i^2 + 1 \text{ for } i = 0, 1, 2 \dots, 10.$$

We note that this is a "collapsing sum".
$(f(1) - f(0)) + (f(2) - f(1)) + (f(3) - f(2))$
$\qquad\qquad\qquad\qquad + (f(10) - f(9)).$
All terms cancel except $-f(0)$ and $f(10)$.
The sum is $f(10) - f(0) = 101 - 1 = 100.$

Area

Concepts to Master

Interval partition; Norm of a partition; Approximation of area under a curve; Area under a curve

Summary and Focus Questions

The goal of this section is to determine the area under a curve $y = f(x)$, above the x-axis, and between lines $x = a$ and $x = b$.

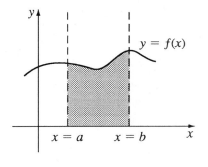

A partition P of $[a, b]$ is a set of subintervals $[x_0, x_1]$, $[x_1, x_2]$, ...$[x_{n-1}, x_n]$, where $a = x_0 < x_1 < x_2 < ... < x_{n-1} < x_n = b$. The width of the ith subinterval is $\Delta x_i = x_i - x_{i-1}$. The norm of P, denoted $\|P\|$, is the largest of all Δx_i.

For example, P: $[2, 4]$, $[4, 5]$, $[5, 8]$, $[8, 9]$ is a partition of $[2, 9]$ and has norm 3 (the width of $[5, 8]$).

A partition of $[a, b]$ into n subintervals of equal widths has norm $\dfrac{b-a}{n}$.

To approximate the area described above:

(1) Partition $[a, b]$ into n subintervals.
(2) Select a point x_i^* in each subinterval.

(3) Compute $\displaystyle\sum_{i=1}^{n} f(x_i^*) \cdot \Delta x_i$.

For some functions the area may be determined exactly by:

(1) Partition $[a, b]$ into subintervals of equal length $\Delta x = \dfrac{b - a}{n}$.

(2) Select a point x_i^* in each subinterval and write it in terms of n and i. Do so consistently. For example, chosing x_i^* to be the left endpoint of $[x_{i-1}, x_i]$ means $x_i^* = a + (i - 1)\Delta x$.

(3) Write the approximating sum $\sum\limits_{i=1}^{n} f(x_i^*) \cdot \Delta x$ as a function of n, as was done in Section 4.1.

(4) Determine the limit of the expression found in step (3) as $n \to \infty$.

---

1) For the interval $[3, 8]$ the norm of the partition $[3, 4]$, $[4, 6]$, $[6, 7]$, $[7, 8]$ is _____.

2 (the length of the widest subinterval).

2) True or False:

The area of a region bounded by $y = f(x)(\geq 0)$, $y = 0$, $x = a$, $x = b$ is the limit of the sums of the areas of approximating rectangles.

True.

3) Using x_i^* as the midpoint of each interval of the partition $[1, 3]$, $[3, 6]$, $[6, 10]$ approximate the area under $f(x) = 1 + 4x^2$ between $x = 1$ and $x = 10$.

Here $n = 3$.

i	subint	Δx_i	x_i^*	$f(x_i^*)$
1	$[1, 3]$	2	2	17
2	$[3, 6]$	3	4.5	82
3	$[6, 10]$	4	8	257

$$\sum_{i=1}^{3} f(x_i^*)\,\Delta x_i = 2(17) + 3(82) + 4(257)$$

$= 1308$. (We shall later see that the exact area is 1341).

4) Find the area under $y = 10 - 2x$ between $x = 1$ and $x = 4$. Take x_i^* to be the right endpoint. Use subintervals of equal width.

With n subintervals we have

$$\Delta x = \frac{b - a}{n} = \frac{4 - 1}{n} = \frac{3}{n}.$$

$$
\begin{array}{c}
\overset{\Delta x}{}\quad\overset{\Delta x}{}\quad\overset{\Delta x}{}\quad\overset{\Delta x}{} \\
[\text{----------}\bullet\text{----------}\bullet\text{----------}\bullet\text{----------}\bullet\text{----------------------}] \\
\underset{= x_0}{1}\qquad x_1 \qquad x_2 \qquad x_3 \qquad\qquad\qquad \underset{= x_n}{4}
\end{array}
$$

$x_1 = 1 + \Delta x$, $x_2 = 1 + 2\Delta x$, and in general $x_i = 1 + i\,\Delta x$. Selecting x_i^* to be the right endpoint means $x_i^* = x_i$.

Thus
$$
\begin{aligned}
x_i^* &= 1 + i\,\Delta x \\
&= 1 + i\left(\frac{3}{n}\right) \\
&= 1 + \frac{3i}{n}.
\end{aligned}
$$

Hence
$$
\begin{aligned}
f(x_i^*) &= 10 - 2\left(1 + \frac{3i}{n}\right) \\
&= 8 - \frac{6i}{n}.
\end{aligned}
$$

The approximating sum is

$$\sum_{i=1}^{n} f(x_i^*) \, \Delta x = \sum_{i=1}^{n} \left(8 - \frac{6i}{n}\right)\left(\frac{3}{n}\right)$$

$$= \sum_{i=1}^{n} \left(\frac{24}{n} - \frac{18i}{n^2}\right)$$

$$= \sum_{i=1}^{n} \frac{24}{n} - \sum_{i=1}^{n} \frac{18i}{n^2}$$

$$= \frac{24}{n} \sum_{i=1}^{n} 1 - \frac{18i}{n^2} \sum_{i=1}^{n} i$$

$$= \frac{24}{n}(n) - \frac{18i}{n^2}\left(\frac{n(n+1)}{2}\right)$$

$$= 24 - 9\frac{n(n+1)}{n^2}.$$

Finally $\lim\limits_{n \to \infty} \left(24 - 9\frac{n(n+1)}{n^2}\right)$

$$= 24 - 9(1) = 13.$$

The Definite Integral

Concepts to Master

Definitions of $\int_a^b f(x)\,dx$, Riemann Sum; Integrability

Summary and Focus Questions

Given a partition P, a <u>Riemann sum</u> for a function f on $[a, b]$ is the number of the form

$$\sum_{i=1}^{n} f(x_i^*)\,\Delta x_i\,,$$

where x_i^* is chosen from $[x_{i-1}, x_i]$, $i = 1, 2, 3 \dots, n$.

The <u>definite integral of $f(x)$ from a to b</u>, denoted $\int_a^b f(x)dx$, is the limit of the values of the Riemann sums for partitions P whose norms approach 0. This means $\int_a^b f(x)dx$ is a number such that for all $\varepsilon > 0$, there exists $\delta > 0$ such that if P is a partition of $[a, b]$ with $\|P\| < \delta$, then $\left| \sum_{i=1}^{n} f(x_i^*)\Delta x_i - \int_a^b f(x)dx \right| < \varepsilon$ for any choice of numbers x_i^* in the subintervals $[x_{i-1}, x_i]$ of P. When this limit exists we say that f is <u>integrable</u> on $[a, b]$.

The definite integral always exists for continuous functions and monotonic functions on closed intervals.

For $f(x) \geq 0$ on $[a, b]$, $\int_a^b f(x)dx$ is the area under the graph of f between $x = a$ and $x = b$. For functions that are not non-negative, the integral represents a kind of "net area".

For example, for the function graphed below $\int_a^b f(x)dx = -A_1 + A_2 - A_3 + A_4$.

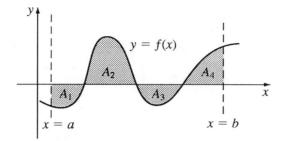

The simplest Riemann sum to calculate is the case where the partition P is regular (all Δx_i are equal to $\dfrac{b-a}{n}$) and $x_i^* = x_i$, the right endpoint of the i-th subinterval. In this case the Riemann sum is

$$\frac{b-a}{n} \sum_{i=1}^{n} f\left(a + i \frac{b-a}{n}\right).$$

Using midpoints for x_i^* and regular partitions, the Riemann sum is

$$\frac{b-a}{n} \sum_{i=1}^{n} f(\overline{x_i}), \text{ where } \overline{x_i} = \frac{x_i + x_{i-1}}{2}.$$

1) Construct a Riemann sum for $f(x) = x^2$ on the interval $[3, 10]$ using the partition $[3, 4], [4, 6], [6, 9], [9, 10]$.

We must select x_i^* points in $[x_{i-1}, x_i]$ and calculate Δx_i for each subinterval:

Select $x_1^* = 3 \in [3, 4]$; $\Delta x_1 = 1$
$x_2^* = 5 \in [4, 6]$; $\Delta x_2 = 2$
$x_3^* = 8 \in [6, 9]$; $\Delta x_3 = 3$
$x_4^* = 9 \in [9, 10]$; $\Delta x_4 = 1$
(Your x_i^* choices may be different.)

The resulting Riemann sum for these choices of x_i^* is:

$$\sum_{i=1}^{4} f(x_i^*)\Delta x_i$$

$$= f(3) \cdot 1 + f(5) \cdot 2 + f(8) \cdot 3 + f(9) \cdot 1$$
$$= (9)1 + (25)2 + (64)3 + (81)1$$
$$= 332.$$

2) Does $\displaystyle\int_1^6 \|x\| \, dx$ exist ?

Yes, because $f(x) = \|x\|$ is increasing (monotone) on [1, 6].

3) Approximate $\displaystyle\int_1^4 x^3 dx$ using a regular partition with $n = 3$ and midpoints for x_i^*.

$$\frac{b-a}{n} = \frac{4-1}{3} = 1.$$

The Riemann sum is calculated as follows:

i	subint	$\overline{x}_i$ mid pt	$f(\overline{x}_i)$
1	[1, 2]	1.5	3.375
2	[2, 3]	2.5	15.625
3	[3, 4]	3.5	42.875

$$\frac{b-a}{n} \sum_{i=1}^{3} f(\overline{x}_i)$$

$$= 1(3.375 + 15.625 + 42.875)$$
$$= 61.875.$$

4) Write a definite integral for the shaded area.

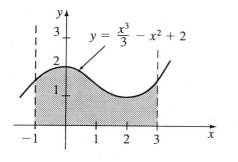

$$\int_{-1}^{3} \left(\frac{x^3}{3} - x^2 + 2 \right) dx.$$

5) Suppose $f(x)$ is continuous and increasing on $[a, b]$. Let $\Re$ be the value of a Riemann sum for $\int_{a}^{b} f(x)\, dx$ using the left endpoint of each subinterval. Is $\Re$ greater than, equal to, or less than $\int_{a}^{b} f(x)\, dx$?

Less than, since the left endpoint will have the smallest value of $f(x)$ in each subinterval.

Properties of the Definite Integral

Concepts to Master

Properties of $\int_a^b f(x)dx$

Summary and Focus Questions

Several integral properties allow definite integrals to be evaluated and estimated. If f and g are integrable functions on an interval containing a, b, and c,

$$\int_a^b f(x)\ dx = - \int_b^a f(x)dx.$$

$$\int_a^b f(x)dx = \int_a^c f(x)dx + \int_c^b f(x)dx.$$

$$\int_a^b (f(x) \pm g(x))\ dx = \int_a^b f(x)dx \pm \int_a^b g(x)dx.$$

$$\int_a^b kf(x)dx = k \int_a^b f(x)dx \text{ (where } k \text{ is a constant).}$$

If $f(x) \geq g(x)$ for all $x \in [a, b]$, then $\int_a^b f(x)dx \geq \int_a^b g(x)dx.$

If $m \leq f(x) \leq M$ for $x \in [a, b]$, then $m(b - a) \leq \int_a^b f(x)dx \leq M(b - a).$

1) What are the missing limits of integration?

$$\int_4^{\square} x^3 dx = \int_{\square}^{\square} x^3 dx + \int_5^7 x^3 dx$$

$$\int_4^7, \int_4^5, \int_5^7 .$$

2) True or False:

a) $\int_0^1 (\sin x + \cos x)dx =$

$\int_0^1 \sin x \, dx + \int_0^1 \cos x \, dx.$

True.

b) $\int_0^1 x(x^2 + 1)dx =$

$\left(\int_0^1 x \, dx\right)\left(\int_0^1 (x^2 + 1) \, dx\right).$

False.

c) $\int_0^1 x(x^2 + 1)dx = x\int_0^1 (x^2 + 1) \, dx.$

False.

d) $\int_0^1 x(t^2 + 1)dt = x \int_0^1 (t^2 + 1) \, dt.$

True, (x is a constant here because t is the variable of integration).

e) $\int_0^{\pi/4} \tan x\, dx \geq 0.$

True, because for $0 \leq x \leq \dfrac{\pi}{4}$, $\tan x \geq 0$.

f) $\int_0^1 2^x dx \leq \int_0^1 x\, dx.$

False, for $0 \leq x \leq 1$, $x < 2^x$, hence

$$\int_0^1 x\, dx \leq \int_0^1 2^x\, dx.$$

g) $\int_1^3 \dfrac{1}{x} dx = \int_1^5 \dfrac{1}{x} dx + \int_5^3 \dfrac{1}{x} dx.$

True.

3) Using $1 \leq x^2 + 1 \leq 10$ for $1 \leq x \leq 3$,

estimate $\int_1^3 (x^2+1)dx.$

Since $1 \leq x^2 + 1 \leq 10$ on $[1, 3]$,

$$1(3-1) \leq \int_1^3 (x^2 + 1)\, dx \leq 10(3-1).$$

Thus $2 \leq \int_1^3 (x^2 + 1)\, dx \leq 20.$

The Fundamental Theorem of Calculus

Concepts to Master

A. Both forms of the Fundamental Theorem of Calculus
B. Evaluation of definite integrals using the Fundamental Theorem; Antiderivative properties

Summary and Focus Questions

A. The Fundamental Theorem of Calculus may be expressed in two different forms:

1. If f is continuous on $[a, b]$ and if g is a function defined by $g(x) = \int_a^x f(t)dt$, then $g'(x) = f(x)$ for all $x \in [a, b]$.

For example, if $g(x) = \int_2^x (t^2 + 2t)dt$, then $g'(x) = x^2 + 2x$.

2. If f is continuous on $[a, b]$ and F is an antiderivative of f, then
$\int_a^b f(x)dx = F(b) - F(a)$.

For example, since an antiderivative of $f(x) = 6x^2 + 2x$ is $F(x) = 2x^3 + x^2$, we have

$\int_1^3 (6x^2 + 2x)dx = F(3) - F(1) = [2(3)^3 + 3^2] - [2(1)^3 + 1^2] = 60$.

1) If $h(x) = \int_4^x \sqrt{t}\, dt$, then $h'(x) = $ _____.

$h'(x) = \sqrt{x}$. (Remember that h is a function of x, not t.)

$\left[\frac{-2\pi}{3} + \frac{1}{81}\right] - \left[\frac{-4}{5} + \frac{1}{3}\right]$

$\frac{-2b}{81} \cdot \frac{54}{81}$

Section 5.5: The Fundamental Theorem of Calculus

2) If $f(x)$ is continuous on $[a, b]$, then for all $x \in [a, b]$, $\dfrac{d}{dx} \displaystyle\int_a^x f(t)\, dt = $ _____.

$f(x)$. This is the first form of the Fundamental Theorem of Calculus.

3) If $f'(x)$ is continuous on $[a, b]$, then for all $x \in [a, b]$, $\displaystyle\int_a^x f'(x)dx = $ _____.

$f(b) - f(a)$. This is the second form of the Fundamental Theorem of Calculus.

4) Evaluate $\displaystyle\int_0^1 (1 - x)dx$.

An antiderivative of $1 - x$ is $F(x) = x - \dfrac{x^2}{x}$.

Thus $\displaystyle\int_0^1 (1 - x)\, dx = F\,(1) - F\,(0)$

$= \left(1 - \dfrac{1}{2}\right) - \left(0 - \dfrac{0}{2}\right) = \dfrac{1}{2}$.

B. The notation for the antiderivative of f is $\int f(x)dx$, the <u>indefinite integral of f with respect to x</u>. For example, $\int 2x\, dx = x^2 + C$, where C is an arbitrary constant, the <u>constant of integration</u>.

Here are some properties of antiderivatives:

The derivative of $\int f(x)\, dx$ is $f(x)$, by definition.

$\displaystyle\int c f(x)\, dx = c \int f(x)\, dx \quad (c, \text{ a constant})$

$\displaystyle\int (f(x) \pm g(x))\, dx = \int f(x)\, dx \pm \int g(x)\, dx$

$\displaystyle\int x^n\, dx = \frac{x^{n+1}}{n + 1} + C \text{ for } n \neq -1 \qquad \int \frac{1}{x}\, dx = \ln |x| + C$

$$\int e^x \, dx = e^x + C \qquad\qquad \int a^x \, dx = \frac{a^x}{\ln a} + C$$

$$\int \sin x \, dx = -\cos x + C \qquad\qquad \int \sec x \, \tan x \, dx = \sec x + C$$

$$\int \cos x \, dx = \sin x + C \qquad\qquad \int \csc^2 x \, dx = -\cot x + C$$

$$\int \sec^2 x \, dx = \tan x + C \qquad\qquad \int \csc x \, \cot x \, dx = -\csc x + C$$

$$\int \frac{1}{x^2 + 1} \, dx = \tan^{-1}x + C \qquad\qquad \int \frac{1}{\sqrt{1 - x^2}} \, dx = \sin^{-1}x + C$$

The second form of the Fundamental Theorem of Calculus provides a handy procedure for evaluating some integrals without resorting to limits of Riemann sums.

1. Find $\int f(x) \, dx = F(x)$ (any antiderivative of f).

2. Compute $F(x) \Big|_a^b = F(b) - F(a)$.

Remember that the method may be applied when f is continous on $[a, b]$.

5) Evaluate

a) $\int x^4 \, dx$

$$\frac{x^5}{5} + C$$

b) $\int (6x + 12x^3) dx$

$$3x^2 + 3x^4 + C$$

c) $\int (2 \cos x - \sec^2 x) dx$

$$2 \sin x - \tan x + C$$

d) $\int \dfrac{1}{2\sqrt{1-x^2}}\, dx$

$\dfrac{1}{2}\sin^{-1}x + C$

6) Determine

a) $\displaystyle\int_0^1 \sqrt{x}\ dx$

$\displaystyle\int_0^1 x^{\frac{1}{2}}dx = \dfrac{x^{\frac{3}{2}}}{\frac{3}{2}}$

$= \dfrac{2}{3}x^{\frac{3}{2}}\Big|_0^1$

$= \dfrac{2}{3}(1^{\frac{3}{2}} - 0^{\frac{3}{2}}) = \dfrac{2}{3}.$

b) $\displaystyle\int_{-1}^3 x^{-3}dx$

The Fundamental Theorem does not apply because x^{-3} is not continuous at $x = 0$. We will have to wait until Chapter 7, Section 9, to evaluate this integral.

c) $\displaystyle\int_{-1}^1 e^x\ dx$

$\displaystyle\int_{-1}^1 e^x\ dx = e^x\Big|_{-1}^1 = e - \dfrac{1}{e}.$

7) True or False:

a) $\int 5 f(x)\ dx = 5 \int f(x)dx.$

True.

b) $\int [f(x)]^2 dx = (\int f(x)dx)^2.$

False.

The Substitution Rule

Concepts to Master

Integral evaluation using substitution

Summary and Focus Questions

The method of substitution is a process of rewriting an integral initially given in the form $\int f(g(x))g'(x)dx$ to the form $\int f(u)du$, where $u = g(x)$. It requires analysing the given integral to determine a proper choice for u and the corresponding differential $du = g'(x)dx$. For example, $\int (x^2+1)^3x \, dx$ has the best choice $u = x^2 + 1$. For then the differential $du = 2x \, dx$ and thus $x \, dx = \dfrac{du}{2}$. Now make the substitutions:

$$\int(x^2 + 1)^3x \, dx = \int u^3 \frac{du}{2} = \frac{1}{2}\int u^3du = \frac{1}{2}\frac{u^4}{4} + C = \frac{(x^2 + 1)^4}{8} + C.$$

Frequently the integral will contain a multiple of $g'(x)dx$ with the final integral containing a constant (like the $\frac{1}{2}$ above).

In the case of a definite integral, such as

$\displaystyle\int_0^1 (x^2 + 1)^3x \, dx$, you may either

1) evaluate using the limits 0 and 2 after substitution:

$$\int_0^2(x^2 + 1)^3x \, dx = \frac{(x^2 + 1)^4}{8} \bigg|_0^2 = \frac{625}{8} - \frac{1}{8} = 78.$$

or 2) change the limits to fit u:

Section 5.6: The Substitution Rule

Since $u(x) = x^2 + 1$, $u(2) = 5$ and $u(0) = 1$.

Hence $\int_0^1 (x^2 + 1)^3 x \, dx = \frac{1}{2}\int_1^5 u^3 du = \frac{u^4}{8}\Big|_1^5 = \frac{625}{8} - \frac{1}{8} = 78.$

1) Use the given substitution to evaluate

$\int x\sqrt{x^2 + 3} \, dx$, $u = x^2 + 3$.

$u = x^2 + 3$, so $du = 2x \, dx$ and $x \, dx = \frac{du}{2}$.

$\int x(x^2 + 3)^{\frac{1}{2}} dx = \int u^{\frac{1}{2}} \frac{du}{2}$

$= \frac{1}{2}\int u^{\frac{1}{2}} du = \frac{1}{2} \cdot \frac{2}{3} u^{\frac{3}{2}}$

$= \frac{1}{3}(x^2 + 3)^{3/2} + C.$

2) State a substitution $u = g(x)$ that will simplify each of the following. Write the simplified integral and evaluate.

a) $\int (x^3 + 3x)^4 (x^2 + 1) dx$

$u = \underline{\hspace{2cm}}$

$du = \underline{\hspace{2cm}}$

$u = x^3 + 3x$

$du = (3x^2 + 3)dx = 3(x^2 + 1)dx.$

Thus $(x^2 + 1)dx = \frac{du}{3}$ and

$\int (x^3 + 3x)^4 (x^2 + 1) dx = \int u^4 \frac{du}{3}$

$= \frac{1}{3}\int u^4 du = \frac{1}{3}\frac{u^5}{5} = \frac{1}{15} u^5$

$$= \frac{1}{15}(x^3 + 3x)^5 + C.$$

b) $\int x^2 \sin x^3 \, dx$

$u =$ _____
$du =$ _____

$u = x^3$

$du = 3x^2 \, dx.$

Thus $x^2 dx = \frac{du}{3}$ and $\int x^2 \sin x^3 dx$

$$= \int \sin u \; \frac{du}{3} = \frac{1}{3} \int \sin u \, du = -\frac{1}{3} \cos u$$

$$= -\frac{1}{3} \cos x^3 + C.$$

c) $\int \tan^2 x \sec^2 x \, dx$

$u =$ _____
$du =$ _____

$u = \tan x$

$du = \sec^2 x \, dx.$

Thus $\int (\tan x)^2 \sec^2 x \, dx = \int u^2 \, du = \frac{1}{3} u^3$

$$= \frac{1}{3} \tan^3 x + C.$$

d) $\int x \, e^{2x^2} \, dx$

$u =$ _____
$du =$ _____

$u = e^{2x^2}$, $du = 4x \, dx.$

Thus $x \, dx = \frac{du}{4}$ and

$$\int x \, e^{\,2x^2} \, dx = \int e^u \, \frac{du}{4} = \frac{1}{4} \int e^u \, du$$

$$= \frac{1}{4} \, e^u = \frac{1}{4} \, e^{2x^2} + C.$$

3) Evaluate $\displaystyle\int_0^1 \sqrt{2x + 1} \, dx.$

$$u = 2x + 1, \quad du = 2dx, \quad dx = \frac{du}{2}.$$

Thus $\displaystyle\int_0^1 (2x + 1)^{1/2} dx = \int u^{1/2} \frac{du}{2}$

$$= \frac{1}{2} \int u^{1/2} du = \frac{1}{2} \cdot \frac{2}{3} u^{3/2}$$

$$= \frac{1}{3} \, (2x + 1) \, \bigg|_0^1 = \frac{1}{3} \, [3 - 1] = \frac{2}{3}.$$

4) Evaluate $\displaystyle\int_0^1 e^x \sin e^x \, dx.$

$$u = e^x, \, du = e^x \, dx$$

$$\int_0^1 e^x \sin e^x \, dx = \int_0^1 \sin u \, du$$

$$= -\cos u = -\cos e^x \, \bigg|_0^1$$

$$= -\cos e^1 - (-\cos e^0)$$

$$= \cos 1 - \cos e.$$

5) Evaluate by changing the limits of integration during substitution:

$$\int_0^{\pi/6} (\cos^3 x + 1) \sin x \, dx$$

$u = \cos x$, $du = -\sin x \, dx$, so

$\sin x \, dx = -du$.

At $x = 0$, $u = \cos 0 = 1$.

At $x = \dfrac{\pi}{6}$, $u = \cos\dfrac{\pi}{6} = \dfrac{1}{2}$.

Therefore $\displaystyle\int_0^{\pi/6} (\cos^3 x + 1) \sin x \, dx$

$$= \int_1^{1/2} (u^3 + 1) \, -du = -\int_1^{1/2} (u^3 + 1) \, du$$

$$= \int_{1/2}^1 (u^3 + 1) \, du = \left(\frac{u^4}{4} + u \right) \Big|_{\frac{1}{2}}^1$$

$$= \left(\frac{1}{4} + 1 \right) - \left(\frac{1}{64} + \frac{1}{2} \right) = \frac{47}{64}.$$

The Logarithm Defined as an Integral

Concepts to Master

Integral definition of $\ln x$; Definition of e, a^x, $\log_a x$

Summary and Focus Questions

Disregarding the definitions and results of Chapter 3, we can start by defining natural log and then define exponentials and logarithms in terms of $\ln x$:

$$\ln x = \int_1^x \frac{1}{t}\, dt \quad \text{for } x > 0$$

e is the unique number for which $\ln e = 1$.

$y = e^x$ is the inverse function of $y = \ln x$.

$a^x = e^{x \ln a}$, for $a > 0$.

$\log_a x = y$ means $a^y = x$.

1) Find an expression for the shaded area:

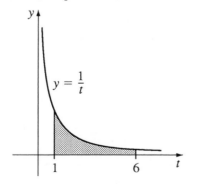

The shaded area is $\int_1^6 \frac{1}{t}\, dt$ which is $\ln 6$.

2) Find the number k on the t-axis such that:

a) the shaded area is 1.

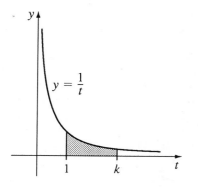

The shaded area is $\int_1^k \frac{1}{t}\, dt = \ln k = 1$, hence $k = e$.

b) the shaded area is 1.

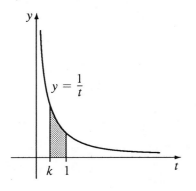

The area $= \int_k^1 \frac{1}{t}\, dt = \ln 1 - \ln k = -\ln k = 1$.

Thus $\ln \frac{1}{k} = 1$, so $\frac{1}{k} = e$ or $k = \frac{1}{e}$.

3) Give three different definitions for the number e.

1. e is the unique number for which $\ln e = 1$.

2. e is the unique number for which $\lim_{h \to 0^+} \frac{e^h - 1}{h} = 1$.

3. $e = \lim_{x \to 0^+} (1 + x)^{1/x}$.

6

Applications of Integration

Area between Curves

Concepts to Master

A. Area between $y = f(x)$, $y = g(x)$, $x = a$, $x = b$
B. Area between $x = f(y)$, $x = g(y)$, $y = c$, $y = d$
C. Area enclosed by two curves

Summary and Focus Questions

A. The area of the region bounded by
continuous $y = f(x)$ and $y = g(x)$
and $x = a$, $x = b$ with $f(x) \geq g(x)$
on $[a, b]$ may be approximated
by a Riemann sum with terms
$[f(x_i^*) - g(x_i^*)]\Delta x_i$.

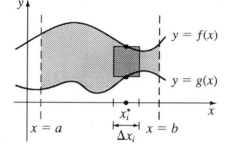

Thus the area is: $A = \displaystyle\lim_{\|P\| \to 0} \sum_{i=1}^{n} [f(x_i^*) - g(x_i^*)]\Delta x_i = \int_a^b [f(x) - g(x)]dx$

1) Set up a definite integral for the area of
the region bounded by $y = x^3$,
$y = 3 - x$, $x = 0$, $x = 1$.

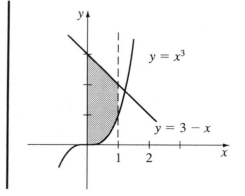

Section 6.1: Area between Curves

For $0 \le x \le 1$, $3 - x \ge x^3$. The area is

$$\int_0^1 [3 - x - x^3]\, dx \quad (= \frac{9}{4}).$$

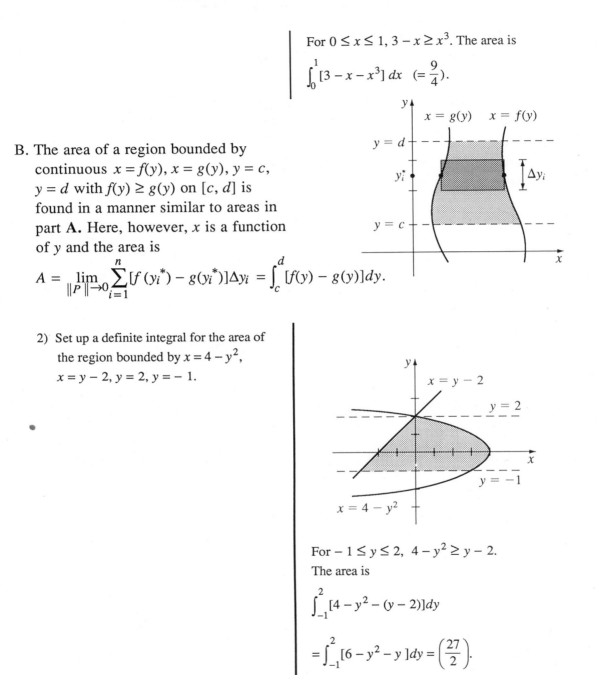

B. The area of a region bounded by
continuous $x = f(y)$, $x = g(y)$, $y = c$,
$y = d$ with $f(y) \ge g(y)$ on $[c, d]$ is
found in a manner similar to areas in
part **A.** Here, however, x is a function
of y and the area is

$$A = \lim_{\|P\| \to 0} \sum_{i=1}^{n} [f(y_i^*) - g(y_i^*)]\Delta y_i = \int_c^d [f(y) - g(y)]\, dy.$$

2) Set up a definite integral for the area of
the region bounded by $x = 4 - y^2$,
$x = y - 2$, $y = 2$, $y = -1$.

For $-1 \le y \le 2$, $4 - y^2 \ge y - 2$.
The area is

$$\int_{-1}^{2} [4 - y^2 - (y - 2)]\, dy$$

$$= \int_{-1}^{2} [6 - y^2 - y]\, dy = \left(\frac{27}{2}\right).$$

Section 6.1: Area between Curves

C. Some regions may be described so that either the $\int ... \, dx$ form or $\int ... \, dy$ form may be used. You should select the form with the easier integral.

To find the area of regions such as the types sketched below, the x coordinates of the points of intersection a, b, c, ... must be found. This is done by setting $f(x) = g(x)$ and solving for x. Often a sketch of the functions will help.

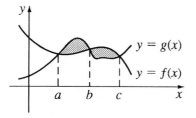

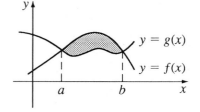

$$\text{Area} = \int_a^b [f(x) - g(x)] \, dx$$

$$\text{Area} = \int_a^b [g(x) - f(x)] \, dx + \int_b^c [f(x) - g(x)] \, dx$$

3) Write an expression involving definite integrals for the area of each shaded region.

Section 6.1: Area between Curves

a)

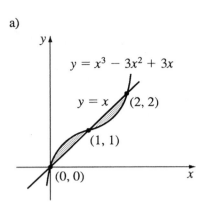

$$\int_0^1 ((x^3 - 3x^2 + 3x) - x)dx$$

$$+ \int_1^2 (x - (x^3 - 3x^2 + 3x))dx$$

$$\left(= \frac{1}{4} + \frac{1}{4} = \frac{1}{2} \right).$$

b)

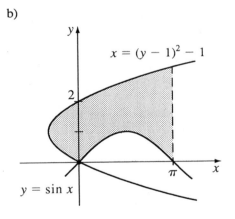

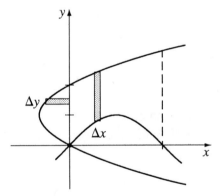

Using the y-axis to divide the region into two parts, the total area is

$$\int_0^2 - ((y - 1)^2 - 1)dy$$

$$+ \int_0^\pi \left((1 + \sqrt{x + 1}) - \sin x \right)dx.$$

(Solving $x = (y - 1)^2 - 1$ for y gives $y = 1 + \sqrt{x + 1}$.)

Section 6.1: Area between Curves

4) Write an expression involving definite integrals for the area of the regions bounded by:

   a) $y = 2x$ and $y = 8 - x^2$

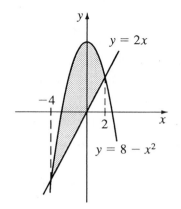

To find the points of intersection set
$2x = 8 - x^2$.
$x^2 + 2x - 8 = 0$
$(x - 2)(x + 4) = 0$, at $x = 2, -4$.

The area is $\int_{-4}^{2} [8 - x^2 - 2x]dx$ $(= 36)$.

   b) $y = x$, $y = -x$, $y = 2x - 3$

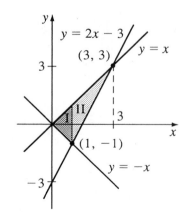

Divide the area into two regions:
Region I is bounded by $x = 0$, $x = 1$, $y = x$ and $y = -x$.
Region II is bounded by $x = 1$, $x = 3$, $y = x$

Section 6.1: Area between Curves

and $y = 2x - 3$.

The area is

$$\int_0^1 [(x - (-x)]dx + \int_1^3 [x - (2x - 3)]\, dx$$

$$= \int_0^1 2x\, dx + \int_1^3 (3 - x)\, dx \quad (= 3).$$

c) $y = \sin x$ and $y = \cos x$ for $0 \le x \le \pi$

$$\sin x = \cos x; \quad \frac{\sin x}{\cos x} = 1; \quad \tan x = 1.$$

So $x = \dfrac{\pi}{4}$ for $0 \le x \le \pi$.

$$\text{Area} = \int_0^{\pi/4} (\cos x - \sin x)\, dx + \int_{\pi/4}^{\pi} (\sin x - \cos x)\, dx$$

$$= ((\sqrt{2} - 1) + (1 + \sqrt{2}) = 2\sqrt{2}).$$

Volumes

Concepts to Master

A. Determining volumes of solids by slicing
B. Determining volumes of solids of revolution (disks)
C. Determining volumes of solids of revolution (washers)

Summary and Focus Questions

A. Suppose the cross-sectional area of a
solid cut by planes perpendicular to an
x-axis is known to be $A(x)$ for each
$x \in [a, b]$. The volume of a typical
slice is approximately $A(x_i^*)\Delta x_i$ and
thus the total volume obtained by this
method of slicing is

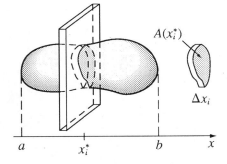

$$V = \lim_{\|P\| \to 0} \sum_{i=1}^{n} A(x_i^*)\, \Delta x_i = \int_a^b A(x)\, dx.$$

In many problems, the x-axis will have to be chosen carefully so that the function
$A(x)$ can be determined.

1) Set up a definite integral for the volume
of a right triangular solid that is 1 m,
1 m, and 2 m on its edges (see the figure).

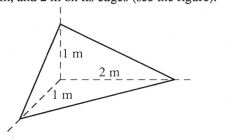

– 183 –

Draw an x-axis along the 2 meter side.

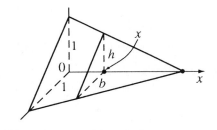

The area of the triangle determined by slicing perpendicular to the x-axis at point x is $A(x) = \frac{1}{2}bh$. By similar triangles,

$$\frac{2-x}{2} = \frac{h}{1}, \text{ so } h = 1 - \frac{x}{2}. \text{ Likewise,}$$

$$b = 1 - \frac{x}{2}, \text{ so } A(x) = \frac{1}{2}\left(1 - \frac{x}{2}\right)^2.$$

The volume is

$$\int_0^2 \frac{1}{2}\left(1 - \frac{x}{2}\right)^2 dx \quad \left(= \frac{1}{3}\right).$$

B. A <u>solid of revolution</u> is a solid obtained by rotating a planar region about a line. For a region bounded by $y = f(x)$, $y = 0$, $x = a$, $x = b$, and rotated about the x-axis, the volume may be approximated by a sum of volumes of disks. The volume of a typical disk is

$\pi \,(\text{radius})^2(\text{thickness}) \approx \pi \,[f(x_i^*)]^2 \Delta x_i.$

Thus the total volume by this <u>disk method</u> is

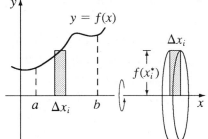

$$V = \lim_{\|P\| \to 0} \sum_{i=1}^{n} \pi \,[f(x_i^*)]^2 \Delta x_i = \int_a^b \pi \,[f(x)]^2 dx.$$

A solid obtained by rotating a region bounded by $x = g(x)$, $x = 0$, $y = c$, $y = d$, about the y-axis has volume

$$V = \int_c^d \pi \, (g(y))^2 dy.$$

2) Set up a definite integral for the volume of each solid of revolution.

   a) Rotate the region $y = 4 - x^2$, $y = 0$, $x = 0$, $x = 2$ about the x-axis.

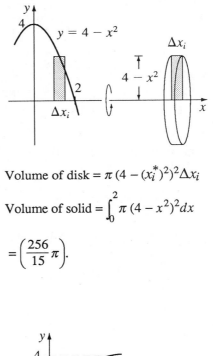

Volume of disk $= \pi \, (4 - (x_i^*)^2)^2 \Delta x_i$

Volume of solid $= \int_0^2 \pi \, (4 - x^2)^2 dx$

$$= \left(\frac{256}{15} \pi \right).$$

   b) The volume of the solid obtained by rotating about the y-axis the region bounded by $y = \sqrt{x}$, $x = 0$, $y = 4$.

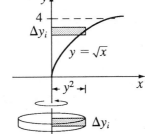

$$\text{Volume of disk} = \pi(y_i^{*2})^2\, \Delta y_i = \pi(y_i^*)^4\, \Delta y$$

$$\text{Volume of solid} = \int_0^4 \pi\, y^4 dy = \left(\frac{1024\pi}{5}\right).$$

C. For a region bounded by $y = f(x)$, $y = g(x)$, (with $f(x) \geq g(x)$), $x = a$, $x = b$, and rotated about the x-axis, the volume of the solid may be approximated by a sum of volumes of washers. The volume of a typical washer is

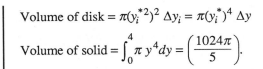

$\pi\,(\text{outer radius})^2(\text{thickness}) - $
$\qquad \pi\,(\text{inner radius})^2(\text{thickness})$
$= \pi\,[(f(x_i^*))^2 - (g(x_i^*))^2]\Delta x_i$

Thus the volume by this <u>washer method</u> is

$$V = \lim_{\|P\|\to 0} \sum_{i=1}^n \pi\,[f(x_i^*)^2 - (g(x_i^*))^2]\Delta x_i = \int_a^b \pi\,[(f(x))^2 - (g(x))^2]dx.$$

Volumes of similar regions rotated about the y-axis are computed using

$$V = \int_c^d \pi\,[(f(y))^2 - (g(y))^2]dy.$$

3) Set up a definite integral for the
volume of the solid obtained by
rotation of the region between
$x = 1$, $x = 2$, $y = x^2$, and $y = x^3$
about the x-axis.

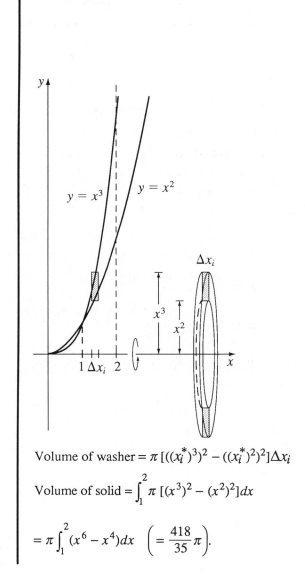

Volume of washer $= \pi\, [((x_i^*)^3)^2 - ((x_i^*)^2)^2]\Delta x_i$

Volume of solid $= \displaystyle\int_1^2 \pi\, [(x^3)^2 - (x^2)^2]dx$

$= \pi \displaystyle\int_1^2 (x^6 - x^4)dx \quad \left(= \frac{418}{35}\pi \right).$

Volumes by Cylindrical Shells

Concepts to Master

Volumes of solids of revolution by the shell method

Summary and Focus Questions

The volume of a region bounded by $y = f(x)$, $y = 0$, $x = a$, $x = b$, and rotated about the y-axis may be approximated by a sum of volumes of shells. The volume of a typical shell is $2\pi(\text{radius})(\text{height})(\text{thickness}) = 2\pi x_i^* f(x_i^*)\Delta x_i$.

Thus the total volume by this <u>method of (cylindrical) shells</u> is

$$V = \lim_{\|P\| \to 0} \sum_{i=1}^{n} 2\pi x_i^* f(x_i^*)\Delta x_i = \int_a^b 2\pi x\, f(x)\ dx.$$

Volumes of solids of revolution about the x-axis are computed similarly.

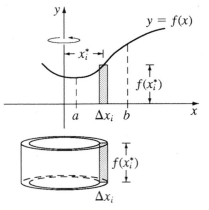

Section 6.3: Volumes by Cylindrical Shells

1) Set up a definite integral for the volume of the solid obtained by rotating about the y-axis the region bounded by $y = \sqrt{x}$, $y = 0$, $x = 1$, $x = 4$.

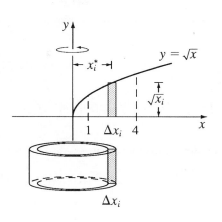

Volume of shell =

$$2\pi x_i^*\left(\sqrt{x_i^*}\right)\Delta x_i = 2\pi(x_i^*)^{3/2}\Delta x_i$$

Volume of solid =

$$\int_1^4 2\pi\, x^{3/2}dx \quad (= \frac{124\pi}{5}).$$

2) Set up a definite integral for the volume of the solid obtained by rotating about the x-axis the region bounded by $x = y^2$, $y = 0$, $x = 4$.

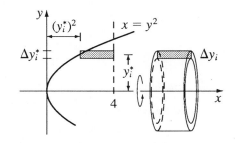

Volume of shell $= 2\pi y_i^*(4 - (y_i^*)^2)\Delta y_i$
$= 2\pi\,(4y_i^* - (y_i^*)^3)\Delta y.$

Volume of solid $= \int_0^2 2\pi\,(4y - y^3)\,dy \quad (= 56\pi).$

Work

Concepts to Master

Work done by a varying amount of force

Summary and Focus Questions

Work is the measure of the total amount of effort, in units of foot-pounds or newton-meters (J, joules). With a constant force F the work done to move an object a distance d is $W = F \cdot d$.

If an object moves along an x-axis from a to b by a varying force $F(x)$, the work done is

$$W = \int_a^b F(x)dx.$$

1) How much work is done lifting an 800 pound piano upward 6 feet?

The force is a constant 800 lbs. and the distance is 6 ft. The work is $W = (800)(6) = 4800$ ft-lb.

2) How much work is done moving a particle along an axis from 0 to $\frac{\pi}{2}$ feet if at each point x the amount of force is $\cos x$ pounds?

$$W = \int_0^{\frac{\pi}{2}} \cos x \, dx = \sin x \, \Big|_0^{\pi/2} = 1 \text{ ft-lb.}$$

3) Find an integral for how much work is done in pumping over the edge all the liquid of density ρ out of a 2 ft. long trough in the shape of an equilateral triangle with sides of 1 ft.

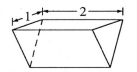

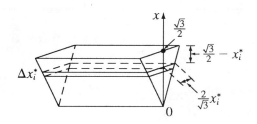

The work done to pump out a slice of the liquid is (density · gravitational constant · volume · height raised). The density is ρ; g is the gravitational constant. The volume of a slice is

$$\left(\frac{2}{3}x_i^*\right)(2)(\Delta x_i) = \frac{4}{\sqrt{3}}x_i^*\Delta x_i.$$

The height raised is $\dfrac{\sqrt{3}}{2} - x_i$.

The work done lifting the slice is

$$\rho g \left(\frac{4}{\sqrt{3}}x_i^*\Delta x_i\right)(\frac{\sqrt{3}}{2} - x_i)$$

The total work is

$$\lim_{\|P\|\to 0}\sum_{i=1}^{n}\rho g \left(\frac{4}{\sqrt{3}}x_i^*\right)\left(\frac{\sqrt{3}}{2} - x_i^*\right)\Delta x_i$$

$$\int_0^{\sqrt{3}/2}\rho g\,\frac{4}{\sqrt{3}}x\left(\frac{\sqrt{3}}{2} - x\right)dx \quad \left(=\frac{\rho g}{4}\right).$$

Average Value of a Function

Concepts to Master

A. Average value of a function
B. Mean Value Theorem for Integrals

Summary and Focus Questions

A. The <u>average value</u> of a continuous function $y = f(x)$ over a closed interval $[a, b]$ is

$$f_{ave} = \frac{1}{b-a} \int_a^b f(x)\, dx.$$

1) Find the average value of $f(x) = 2x + 6x^2$ over $[1, 4]$.

$$f_{ave} = \frac{1}{4-1} \int_1^4 (2x + 6x^2)\, dx$$

$$= \frac{1}{3} (x^2 + 2x^3) \Big|_0^4 = 47.$$

2) Find the average value of $\cos x$ over $[0, \pi]$.

$$\frac{1}{\pi - 0} \int_0^\pi \cos x\, dx = \frac{1}{\pi} \sin x \Big|_0^\pi$$

$$= \frac{1}{\pi} (0 - 0) = 0.\ \text{Intuitively this makes sense}$$

because the "half" of the cosine curve

between $\frac{\pi}{2}$ and π is below the x-axis and

is symmetric with the other half above the

x-axis.

B. The Mean Value Theorem for Integrals:

If f is continuous on a closed interval $[a, b]$ there exists $c \in [a, b]$ such that

$$\int_a^b f(x) \, dx = f(c) \, (b - a).$$

3) Mark on the graph a number c guaranteed
by the Mean Value Theorem for $f(x)$:

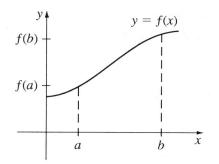

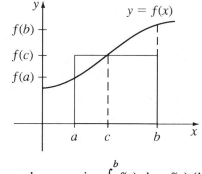

The number c requires $\int_a^b f(x) \, dx = f(c) \, (b - a)$

so the area under $f(x)$ must equal the area of the
rectangle with width $b - a$ and height $f(c)$.

Section 6.5: Average Value of a Function

4) Let $F'(x) = f(x)$ be continuous on $[a, b]$. State the conclusion of the Mean Value Theorem for Integrals for f in terms of the function $F(x)$. Use the Fundamental Theorem of Calculus.

The conclusion is "there is a number c such that $\int_a^b f(x)\, dx = f(c)\,(b-a)$."

By the Fundamental Theorem
$\int_a^b f(x)dx = F(b) - F(a)$. Also

$f(c) = F'(c)$. So the conclusion, in terms of $F(x)$, becomes "there is a number $c \in (a, b)$ such that $F(b) - F(a) = F'(c)(b-a)$" – precisely the Mean Value Theorem of Section 4.2 !

7

Techniques of Integration

"I THINK YOU SHOULD BE MORE EXPLICIT HERE IN STEP TWO."

Integration by Parts

Concepts to Master

Evaluation of <u>integrals by parts</u>

Summary and Focus Questions

The <u>integration by parts</u> formula

$$\int u\, dv = u\, v - \int v\, du$$

comes from the product rule for derivatives: $(uv)' = uv' + u'v$. If used correctly it will replace the integral $\int u\, dv$ with a somewhat easier $\int v\, du$.

The key to its successful use is to identify an integrand as $u\, dv\ (= u\, v'\, dx)$ and check to see if $\int v\, du$ is indeed simpler.

1) Give an appropriate choice for u and dv in each by parts integral and evaluate:

a) $\int x^2 \ln x\, dx$

$u = $ _____

$dv = $ _____

$u = \ln x, \ \ du = \dfrac{1}{x}\, dx$

$dv = x^2 dx, \ \ v = \dfrac{x^3}{3}.$

Thus $\int x^2 \ln x\, dx$

$$= (\ln x)\, \frac{x^3}{3} - \int \frac{x^3}{3} \cdot \frac{1}{x}\, dx$$

$$= \frac{x^3}{3}\, \ln x - \frac{1}{3} \int x^2 dx$$

$$= \frac{x^3}{3}\, \ln x - \frac{1}{9} x^3 + C.$$

b) $\int x \cos x\, dx$

$u = $ _____

$dv = $ _____

$u = x,\ \ du = dx$
$dv = \cos x,\ \ v = \sin x.$

Then $\int x \cos x\, dx$

$$= x \sin x - \int \sin x\, dx$$

$$= x \sin x - (-\cos x)$$
$$= x \sin x + \cos x + C.$$

2) Evaluate $\int x^2 e^{5x}\, dx.$

Let $u = x^2,\ \ dv = e^{5x} dx,$

$du = 2x\, dx,\ \ v = \frac{1}{5} e^{5x}.$

Thus $\int x^2 e^{5x}\, dx$

$$= x^2 \left(\frac{1}{5} e^{5x} \right) - \int \frac{1}{5} e^{5x}\, 2x\, dx$$

$$= \frac{1}{5} x^2 e^{5x} - \frac{2}{5} \int x e^{5x}\, dx.$$

The last integral is easier than the first so
we are on the right track. Apply by parts to it:

Section 7.1 : Integration by Parts

$$\int x\, e^{5x}\, dx = \begin{pmatrix} u = x, \ \ dv = e^{5x}\, dx \\ du = dx, \ \ v = \frac{1}{5}\, e^{5x} \end{pmatrix}$$

$$= x\left(\frac{1}{5}e^{5x}\right) - \int \frac{1}{5}e^{5x}\, dx$$

$$= \frac{1}{5}\, x\, e^{5x} - \frac{1}{25}\, e^{5x}.$$

So the original integral evaluates to

$$\frac{1}{5}\, x^2\, e^{5x} - \frac{1}{5}\, x\, e^{5x} + \frac{1}{25}\, e^{5x} + C.$$

3. Evaluate $\int \sin x \cos x\, dx$ by parts.

$$\int \sin x \cos x\, dx$$

$$\begin{pmatrix} u = \sin x & dv = \cos x\, dx \\ du = \cos x\, dx & v = \sin x \end{pmatrix}$$

$$= (\sin x)(\sin x) - \int (\sin x) \cos x\, dx.$$

Thus $\int \sin x \cos x\, dx$

$$= \sin^2 x - \int \sin x \cos x\, dx.$$

Hence $2 \int \sin x \cos x\, dx = \sin^2 x,$

so $\int \sin x \cos x\, dx = \frac{1}{2}\, \sin^2 x + C.$

Note: This could have been done with the simple substitution $u = \sin x$, $du = \cos x\, dx$

to yield $\int u\, du$. We could also have solved

this with the identity $\sin x \cos x = \frac{1}{2}\, \sin 2x$

and the substitution $u = 2x$.

Trigonometric Integrals

Concepts to Master

Evaluation of integrals using trigonometric identities

Summary and Focus Questions

In this section integrals with certain combinations of trigonometric functions are rewritten using trigonometric identities. The resulting integrals will be easier to evaluate. A summary of the techniques is given in the table on the following page. Note that n and m must be nonnegative integers. If the condition(s) hold, the original integral may be rewritten as indicated.

Original Integral	Condition(s)	Identity to Use	Rewritten Form
	$m =$ odd	$\sin^2 x = 1 - \cos^2 x$	$\int (1 - \cos^2 x)^{(m-1)/2} \cos^n x \, \sin x \, dx$
$\int \sin^m x \, \cos^n x \, dx$	$n =$ odd	$\cos^2 x = 1 - \sin^2 x$	$\int (1 - \sin^2 x)^{(n-1)/2} \sin^m x \, \cos x \, dx$
	m, n even	$\cos^2 x = 1 - \sin^2 x$	$\int (1 - \sin^2 x)^{n/2} \sin^m x \, dx$
	$m =$ odd	$\tan^2 x = \sec^2 x - 1$	$\int (\sec^2 x - 1)^{(m-1)/2} \sec^{n-1} x \, \tan x \, \sec x \, dx$
$\int \tan^m x \, \sec^n x \, dx$	$n =$ even	$\sec^2 x = \tan^2 x + 1$	$\int \tan^m x \, (\tan^2 x + 1)^{(n-2)/2} \sec^2 x \, dx$
	$m =$ even, $n =$ odd	$\tan^2 x = \sec^2 x - 1$	$\int (\sec^2 x - 1)^{m/2} \sec^n x \, dx$
$\int \sin mx \, \sin nx \, dx$		$\sin A \, \sin B = \frac{1}{2}(\cos(A - B) - \cos(A + B))$	$\int \frac{1}{2}(\cos(m - n)x - \cos(m + n)x) dx$
$\int \cos mx \, \cos nx \, dx$		$\cos A \cos B = \frac{1}{2}(\cos(A - B) + \cos(A + B))$	$\int \frac{1}{2}(\cos(m - n)x + \cos(m + n)x) dx$
$\int \sin mx \, \cos nx \, dx$		$\sin A \cos B = \frac{1}{2}(\sin(A - B) + \sin(A + B))$	$\int \frac{1}{2}(\sin(m + n)x + \sin(m - n)x) dx$

Integrals containing $\cot^m x \, \csc^n x$ are handled similarly to $\int \tan^m x \, \sec^n x \, dx$ using the identity $\cot^2 x = \csc^2 x - 1$.

1) Evaluate $\int \sin^5 x \, dx$.

This fits the form $\int \sin^m x \, \cos^n x \, dx$

with $m = 5$, $n = 0$.

$\int \sin^5 x \, dx = \int \sin^4 x \, \sin x \, dx$

$= \int (1 - \cos^2 x)^2 \, \sin x \, dx$

$(u = \cos x, \, du = -\sin x \, dx)$

$$= -\int (1 - u^2)^2 \, du = -\int (1 - 2u^2 + u^4) du$$

$$= -u + \frac{2}{3}u^3 - \frac{1}{5}u^5$$

$$= -\cos x + \frac{2}{3}\cos^3 x - \frac{1}{5}\cos^5 x + C.$$

2) Evaluate $\int \tan x \sec^4 x \, dx$.

Here $m = 1$ and $n = 4$ so we have our choice of identities. We select the $n = $ even case.

$\int \tan x \sec^4 x \, dx$

$= \int \tan x \, (\sec^2 x) \sec^2 x \, dx$

$= \int \tan x \, (\tan^2 x + 1) \sec^2 x \, dx$

$= \int (\tan^3 x + \tan x) \sec^2 x \, dx$

$\begin{pmatrix} u = \tan x \\ du = \sec^2 x \, dx \end{pmatrix}$

$= \int (u^3 + u) du = \dfrac{u^4}{4} + \dfrac{u^2}{2}$

$= \dfrac{1}{4} \tan^4 x + \dfrac{1}{2} \tan^2 x + C.$

3) Evaluate $\int \sin 6x \cos 2x \, dx$.

$\int \sin 6x \cos 2x \, dx$

$= \int \dfrac{1}{2}(\sin 4x + \sin 8x) dx$

$= \dfrac{1}{2} \int \sin 4x \, dx + \dfrac{1}{2} \int \sin 8x \, dx$

$= \dfrac{1}{2}\left(\dfrac{-\cos 4x}{4}\right) + \dfrac{1}{2}\left(\dfrac{-\cos 8x}{8}\right)$

$= -\dfrac{1}{8} \cos 4x - \dfrac{1}{16} \cos 8x + C.$

4) Evaluate $\int \cos^4 x \, dx$.

$$\int \cos^4 x \, dx = \int (\cos^2 x)^2 \, dx$$

$$= \int \left(\frac{1 + \cos 2x}{2} \right)^2 dx$$

$$= \frac{1}{4} \int (1 + 2 \cos 2x + \cos^2 2x) dx$$

$$= \frac{1}{4} \int \left(1 + 2 \cos 2x + \frac{1 + \cos 4x}{2} \right) dx$$

$$= \frac{1}{4} \int \left(\frac{3}{2} + 2 \cos 2x + \frac{1}{2} \cos 4x \right) dx$$

$$= \frac{1}{4} \left[\frac{3}{2} x + \sin 2x + \frac{1}{2} \frac{\sin 4x}{4} \right]$$

$$= \frac{3}{8} x + \frac{1}{4} \sin 2x + \frac{1}{32} \sin 4x + C.$$

5) Evaluate $\int \tan^2 x \sec x \, dx$.

$$\int \tan^2 x \sec x \, dx = \int (\sec^2 x - 1) \sec x \, dx$$

$$= \int (\sec^3 x - \sec x) dx$$

$$= \int \sec^3 x \, dx - \int \sec x \, dx$$

$$= \frac{1}{2} (\sec x \tan x + \ln|\sec x + \tan x|)$$

$$- \ln|\sec x + \tan x| + C$$

$$\left(\int \sec^3 x \, dx \text{ is evaluated by parts} \right)$$

$$= \frac{1}{2} (\sec x \tan x - \ln|\sec x + \tan x|) + C.$$

Trigonometric Substitution

Concepts to Master

Inverse substitution, trigonometric substitution for integrals containing powers of $\sqrt{a^2 - x^2}$, $\sqrt{a^2 + x^2}$, $\sqrt{x^2 - a^2}$

Summary and Focus Questions

The process of <u>inverse substitution</u> for evaluating $\int f(x)\, dx$ replaces x (and dx) by a function of another variable: t, θ, ... The result, possibly after some simplification, may be an integral easier to evaluate.

This section considers inverse substitutions of the form $x = a \sin \theta$, $x = a \tan \theta$, and $x = a \sec \theta$. The resulting integrals are then evaluated by the techniques of Section 7.2. The table on the next page summarizes the substitutions.

A mnemonic triangle may be drawn to help determine the correct trigonometric substitution. Remember that the triangle is drawn for $0 \le \theta \le \dfrac{\pi}{2}$ only and that relationships should be worked out using trigonometric identities $\sin^2\theta + \cos^2\theta = 1$ and $\tan^2\theta = \sec^2\theta - 1$.

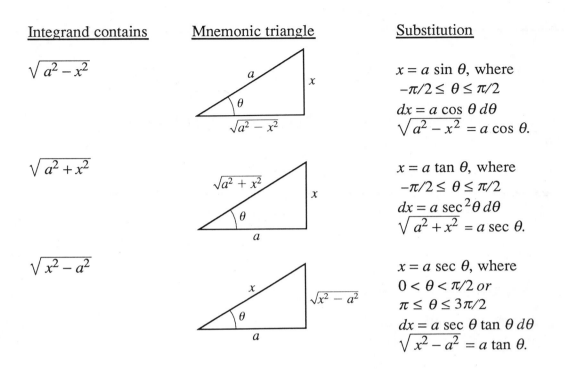

Integrand contains	Mnemonic triangle	Substitution

$\sqrt{a^2 - x^2}$

$x = a \sin \theta$, where
$-\pi/2 \le \theta \le \pi/2$
$dx = a \cos \theta \, d\theta$
$\sqrt{a^2 - x^2} = a \cos \theta$.

$\sqrt{a^2 + x^2}$

$x = a \tan \theta$, where
$-\pi/2 \le \theta \le \pi/2$
$dx = a \sec^2 \theta \, d\theta$
$\sqrt{a^2 + x^2} = a \sec \theta$.

$\sqrt{x^2 - a^2}$

$x = a \sec \theta$, where
$0 < \theta < \pi/2$ or
$\pi \le \theta \le 3\pi/2$
$dx = a \sec \theta \tan \theta \, d\theta$
$\sqrt{x^2 - a^2} = a \tan \theta$.

It may be necessary to algebraically transform an integrand before substitution. One technique is completing the square:

<u>Example:</u>   Evaluate $\displaystyle\int \frac{dx}{\sqrt{x^2 + 4x + 5}}$.

The integrand $\dfrac{1}{\sqrt{x^2 + 4x + 5}}$ does not apparently meet the form $\dfrac{1}{\sqrt{1 + u^2}}$, but

will do so after completing the square:

$$x^2 + 4x + 5 = x^2 + 4x + 4 + 5 - 4 = (x + 2)^2 + 1.$$

Thus $\displaystyle\int \frac{1}{\sqrt{x^2 + 4x + 5}} \, dx = \int \frac{1}{1 + (x + 2)^2}.$

Let $u = x + 2$. Then $du = dx$ and $\displaystyle\int \frac{1}{\sqrt{1 + (x + 2)^2}} = \int \frac{1}{\sqrt{1 + u^2}} du.$

Section 7.3: Trigonometric Substitution

We now use a trigonometric substitution:

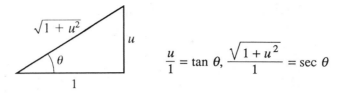

$$\frac{u}{1} = \tan\theta, \quad \frac{\sqrt{1+u^2}}{1} = \sec\theta$$

Let $u = \tan\theta$, $du = \sec^2\theta\, d\theta$, $\sqrt{1+u^2} = \sec\theta$.

Then $\displaystyle\int \frac{1}{\sqrt{1+u^2}}\, du = \int \frac{1}{\sec\theta}\sec^2\theta\, d\theta = \int \sec\theta\, d\theta$

$\qquad = \ln\left|\sec\theta + \tan\theta\right| = \ln\left|\sqrt{1+u^2} + u\right|$

$\qquad = \ln\left|\sqrt{x^2 + 4x + 5} + x + 2\right| + C.$

1) Rewrite each of the following using a trigonometric substitution:

a) $\displaystyle\int x^2\sqrt{4 - x^2}\, dx$

$$\frac{x}{2} = \sin\theta$$

$$\frac{\sqrt{4-x^2}}{2} = \cos\theta$$

Let $x = 2\sin\theta$,
$dx = 2\cos\theta\, d\theta$,
$\sqrt{4 - x^2} = 2\cos\theta$.
Then $\displaystyle\int x^2\sqrt{4 - x^2}\, dx$

$\displaystyle = \int (2\sin\theta)^2\, (2\cos\theta)\,(2\cos\theta\, d\theta)$

$\displaystyle = 16 \int \sin^2\theta\, \cos^2\theta\, d\theta.$

b) $\int x^2 \sqrt{x^2 - 4}\, dx$

$$\frac{x}{2} = \sec\theta$$

$$\frac{\sqrt{x^2 - 4}}{2} = \tan\theta$$

Let $x = 2\sec\theta$,
$dx = 2\sec\theta\tan\theta\, d\theta$,
$\sqrt{x^2 - 4} = 2\tan\theta$.

Then $\int x^2 \sqrt{x^2 - 4}\, dx$

$= \int (2\sec\theta)^2\, (2\tan\theta)\, (2\sec\theta\tan\theta\, d\theta)$

$= 16 \int \sec^3\theta \tan^2\theta\, d\theta$.

c) $\int \frac{x + 1}{\sqrt{4 + x^2}}\, dx$

$$\frac{x}{2} = \tan\theta$$

$$\frac{\sqrt{4 + x^2}}{2} = \sec\theta$$

Let $x = 2\tan\theta$, $dx = 2\sec^2\theta\, d\theta$,
$\sqrt{4 + x^2} = 2\sec\theta$.

Then $\int \frac{x + 1}{\sqrt{4 + x^2}}\, dx$

$= \int \frac{2\tan\theta + 1}{2\sec\theta} \cdot 2\sec^2\theta\, d\theta$

$$= \int (2 \tan \theta + 1) \sec \theta \, d\theta$$

$$= \int 2 \tan \theta \sec \theta \, d\theta + \int \sec \theta \, d\theta.$$

3) Rewrite the following so that a trigonometric substitution may be used

$$\int \frac{x^2}{\sqrt{13 - 6x + x^2}} \, dx$$

$\sqrt{13 - 6x + x^2}$ does not look friendly until we complete the square:
$$13 - 6x + x^2 = 13 + (x^2 - 6x + 9 - 9)$$
$$= x^2 - 6x + 9 + 4 = (x - 3)^2 + 4.$$
Let $u = x - 3$, $du = dx$,
$$x = u + 3 \text{ and } 13 - 6x + x^2 = u^2 + 4$$

Then $\int \dfrac{x^2}{\sqrt{13 - 6x + x^2}} \, dx$

$$= \int \frac{(u + 3)^2}{\sqrt{u^2 + 1}} \, du.$$

The integral is now ready for the substitution $u = \tan \theta$.

Integration of Rational Functions by Partial Fractions

Concepts to Master

Partial fraction form; Determining term coefficients

Summary and Focus Questions

A <u>proper</u> rational function is a function of the form $\dfrac{P(x)}{Q(x)}$ where $P(x)$ and $Q(x)$ are polynomials with the degree of $P(x)$ less than the degree of $Q(x)$. Any rational function may be written, using long division if necessary, as a polynomial plus a proper rational function.

The <u>method of partial fractions</u> for evaluating integrals is all algebra. Its goal is to write a proper $\dfrac{P(x)}{Q(x)}$ integrand as a sum of rational functions each of which you already know how to integrate.

<u>Step 1.</u> Write $Q(x)$ in factored form as a product of powers of <u>distinct</u> linear and irreducible quadratics.

<u>Step 2.</u> Write $\dfrac{P(x)}{Q(x)} = T_1 + T_2 + \ldots + T_n$

where the T_i are terms corresponding to the distinct factors of $Q(x)$ found in Step 1. The forms of T_i are given in the table on the next page.

Section 7.4: Integration of Rational Functions by Partial Fractions

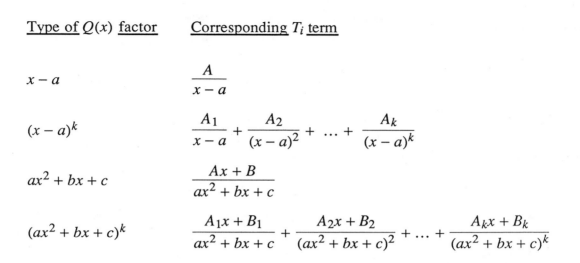

Type of $Q(x)$ factor	Corresponding T_i term
$x - a$	$\dfrac{A}{x - a}$
$(x - a)^k$	$\dfrac{A_1}{x - a} + \dfrac{A_2}{(x - a)^2} + \dots + \dfrac{A_k}{(x - a)^k}$
$ax^2 + bx + c$	$\dfrac{Ax + B}{ax^2 + bx + c}$
$(ax^2 + bx + c)^k$	$\dfrac{A_1 x + B_1}{ax^2 + bx + c} + \dfrac{A_2 x + B_2}{(ax^2 + bx + c)^2} + \dots + \dfrac{A_k x + B_k}{(ax^2 + bx + c)^k}$

Step 3. Write $P(x) = Q(x)[T_1 + T_2 + \dots + T_n]$. Then multiply out the right side, combining like terms.

Step 4. Equate coefficients of like terms in the equation in Step 3. This gives a system of linear equations.

Step 5. Solve the system in Step 4. This is frequently done by solving for one variable in one equation and using this to eliminate that variable in the other equations.

Shortcut: Sometimes, some or all of the unknown coefficients can be found after Step 2 by judicious substitution of x values. The rest of the coefficients may be found by continuing with Steps 3, 4, and 5.

Here is a complete example for the integral

$$\int \frac{4x^3 - 4x^2 + 5x + 1}{(x^2 - 2x + 1)(x^2 + x + 1)}\, dx.$$

The integrand is proper because the degree of the numerator is less than the degree

Section 7.4: Integration of Rational Functions by Partial Fractions

of the denominator.

Step 1: $Q(x) = (x^2 - 2x + 1)(x^2 + x + 1) = (x - 1)^2(x^2 + x + 1)$. Note that $x^2 + x + 1$ is irreducible.

Step 2: $\dfrac{4x^3 - 4x^2 + 5x + 1}{(x - 1)^2(x^2 + x + 1)} = \dfrac{A}{(x - 1)} + \dfrac{B}{(x - 1)^2} + \dfrac{Cx + D}{x^2 + x + 1}$.

Step 3: Multiply by the denominator $Q(x)$ and employ cancellation:

$$4x^3 - 4x^2 + 5x + 1 = A(x - 1)(x^2 + x + 1) + B(x^2 + x + 1) + (Cx + D)(x - 1)^2.$$

Multiply out the right side:
$4x^3 - 4x^2 + 5x + 1 =$
$Ax^3 - A + Bx^2 + Bx + B + Cx^3 - 2Cx^2 + Cx + Dx^2 - 2Dx + D$
and collect like terms:
$4x^3 - 4x^2 + 5x + 1 =$
$(A + C)x^3 + (B - 2C + D)x^2 + (B - 2C + D)x^2 + (B + C - 2D)x + (-A + B + D).$

Step 4: Equate coefficients:
$$4 = A + C$$
$$-4 = B - 2C + D$$
$$5 = B + C - 2D$$
$$1 = -A + B + D$$

Step 5: Solve the system. From the first equation $A = 4 - C$ and thus the others become:

$$
\begin{array}{ccc}
-4 = B - 2C + D & & -4 = B - 2C + D \\
5 = B + C - 2D & \quad or \quad & 5 = B + C - 2D \\
1 = -(4 - C) + B + D & & 5 = B + C + D
\end{array}
$$

Subtracting the last two equations yields $0 = 3D$ or $D = 0$. So the system is
$$-4 = B - 2C$$
$$5 = B + C.$$

Again subtracting, $-9 = -3C$, $C = 3$. Thus $B = 2$ and $A = 1$.

Therefore $\dfrac{4x^3 - 4x^2 + 5x + 1}{(x^2 - 2x + 1)(x^2 + x + 1)} = \dfrac{1}{x - 1} + \dfrac{2}{(x - 1)^2} + \dfrac{3x}{x^2 + x + 1}$.

Finally, the desired integral may be written as

$$\int \frac{1}{x - 1}\, dx + \int \frac{2}{(x - 1)^2}\, dx + \int \frac{3x}{x^2 + x + 1}\, dx \qquad \text{(Whew!)}$$

(In this example we could have first found $B = 2$ by setting $x = 1$ in Step 3. The rest of the procedure would be followed to find A, C, and D.)

1) Write the partial fraction form (Step 2)
   for each

   a) $\dfrac{2x + 1}{x(x + 8)}$

   $\dfrac{A}{x} + \dfrac{B}{x + 8}$

   b) $\dfrac{4x^2 + 11}{(x + 1)^2(x^2 - 3x + 5)}$

   $\dfrac{A}{x + 1} + \dfrac{B}{(x + 1)^2} + \dfrac{Cx + D}{x^2 - 3x + 5}$

   c) $\dfrac{3}{x(x^2 + 2x + 2)^2}$

   $\dfrac{A}{x} + \dfrac{Bx + C}{x^2 + 2x + 2} + \dfrac{Dx + E}{(x^2 + 2x + 2)^2}$

2) Evaluate $\int \dfrac{6-x}{x(x+3)}\, dx$.

(1) $Q(x) = x(x+3)$

(2) $\dfrac{6-x}{x(x+3)} = \dfrac{A}{x} + \dfrac{B}{x+3}$

(3) Clear fractions:

$6 - x = A(x+3) + Bx$

Combine terms:

$6 - x = (A+B)x + 3A$

(4) Equate coefficients:

$A + B = -1$

$3A = 6$

(5) Solving, we see $A = 2$ from the second equation. Thus $2 + B = -1$, $B = -3$.

Therefore $\int \dfrac{6-x}{x(x+3)}\, dx = \int \left(\dfrac{2}{x} + \dfrac{-3}{x+3} \right) dx$

$\qquad = 2 \ln|x| - 3 \ln|x+3| + C.$

Note: At Step 3, after clearing fractions to get

$6 - x = A(x+3) + Bx$

we could substitute $x = 0$ to get

$6 = A(0+3) + 0$, so $A = 2$,

and we could substitute $x = -3$ to get

$6 - (-3) = A(0) + B(-3)$

$9 = -3B$, $B = -3$.

Sometimes, substituting carefully selected values of x can shorten the work.

Rationalizing Substitutions

Concepts to Master

A. Substitution of $u = \sqrt[n]{g(x)}$

B. Substitution of $t = \tan \dfrac{x}{2}$

Summary and Focus Questions

A. In some cases the substitution $u = \sqrt[n]{g(x)}$ in the form $u^n = g(x)$ will transform an integrand involving radicals into a rational function. Such is the case with

$$\int \frac{x}{\sqrt{x-2}} \, dx.$$

Let $u = \sqrt{x-2}$, $u^2 = x - 2$, $x = u^2 + 2$, $dx = 2u \, du$.

Then $\displaystyle \int \frac{x}{\sqrt{x-2}} dx = \int \frac{u^2 + 2}{u} \, 2u \, du = \int (u^2 + 2)du = \frac{u^3}{3} + 2u =$

$$\frac{1}{3}(x-2)^{3/2} + 2\sqrt{x-2} + C.$$

1) Evaluate $\displaystyle \int \frac{\sqrt[3]{x}}{\sqrt[3]{x} + 1} dx$

Let $u = \sqrt[3]{x}$, $x = u^3$, $dx = 3u^2 du$.

The integral becomes

$$\int \frac{u}{u+1} 2u^2 du = 2 \int \frac{u^3}{u+1} \, du.$$

By long division $\dfrac{u^3}{u+1} = u^2 - u + 1 + \dfrac{1}{u+1}.$

$$\int \frac{u^3}{u+1} du = \int \left(u^2 - u + 1 - \frac{1}{u+1} \right) du$$

$$= \frac{u^3}{3} - \frac{u^2}{2} + u - \ln|u+1|.$$

So the original integral is

$$2\left[\frac{x}{3} - \frac{x^{2/3}}{2} + \sqrt[3]{x} - \ln\left|\sqrt[3]{x} + 1\right| \right] + C.$$

2) What substitution should be made for

$$\int \frac{\sqrt[3]{x}}{1 + \sqrt[4]{x}} \, dx?$$

$u = \sqrt[12]{x}$, because then $\sqrt[4]{x}$ $(= u^3)$ and

$\sqrt[3]{x}$ $(= u^4)$ are both integer powers of u.

(*Note*: 12 is the least common multiple of 3 and 4.)

B. Rational functions of $\sin x$ and $\cos x$ can be converted into ordinary rational

functions using the substitution $t = \tan \dfrac{x}{2}$. This results in

$$\sin x = \frac{2t}{1 + t^2}, \quad \cos x = \frac{1 - t^2}{1 + t^2}, \quad dx = \frac{2}{1 + t^2} dt$$

3) Evaluate, using the substitution $t = \tan \dfrac{x}{2}$:

$$\int \frac{\sin x}{\cos x + 1} dx$$

$$\int \frac{\sin x}{\cos x + 1} dx$$

$$= \int \frac{\dfrac{2t}{1 + t^2}}{\dfrac{1 - t^2}{1 + t^2} + 1} \frac{2}{1 + t^2} dt$$

$$= \int \frac{2t}{1 - t^2 + 1 + t^2} \frac{2}{1 + t^2} dt$$

$$= \int \frac{2t}{1 + t^2} dt = \ln \left| 1 + t^2 \right|$$

$$= \ln \left(1 + \tan^2 \frac{x}{2} \right) + C.$$

4) Write the integral of the rational function resulting from the substitution $t = \tan \dfrac{x}{2}$ in

$$\int (1 + \sin^2 x - \cos^2 x)dx.$$

The substitution yields

$$\int \left[1 + \left(\frac{2t}{1 + t^2} \right)^2 - \left(\frac{1 - t^2}{1 + t^2} \right)^2 \right] \frac{2}{1 + t^2} dt$$

which after some algebra is

$$\int \frac{16t^2}{(1 + t^2)^3} dt.$$

Strategy for Integration

Concepts to Master

A. No new concepts about integration; just putting together all the integration techniques

B. Nonintegrable functions

Summary and Focus Questions

A. The two main techniques for evaluating integrals are: <u>substitution</u> and <u>integration by parts</u>. Substitution can take many forms:

<u>Method</u>	<u>Sample</u>
Straight substitution	$\int \dfrac{x^2}{x^3 + 1} dx, \qquad u = x^3 + 1, \ldots$
Trigonometric	$\int \sin^3 x \cos^2 x \, dx, \quad \int \sin^2 x \cos^2 x \sin x \, dx = \ldots$
Trigonometric substitution	$\int \dfrac{x^2}{\sqrt{4 - x^2}} dx, \quad x = 2 \sin \theta, \ldots$
Partial fractions	$\int \dfrac{1}{(x + 2)(x - 3)} dx = \int \left(\dfrac{-1}{x + 2} + \dfrac{1}{x - 3} \right) dx = \ldots$
Rationalizing substitution	$\int \dfrac{x}{\sqrt{x} + 1} dx, \quad u = \sqrt{x}, u^2 = x, \ldots$

Remember your algebra and trigonometric identities and know the integrals in the Table of Integration formulas (7.27) of the text.

1) Indicate what technique should be
   used to begin solving each:

a) $\int \dfrac{e^x}{e^x - 1}$

Straight substitution, $u = e^x - 1$.

b) $\int x^2 \sqrt{16 + x^2} \, dx$

Trigonometric substitution, $x = 4 \tan \theta$.

c) $\int \dfrac{2x + 1}{x^3(x + 1)}$

Partial fractions, $\dfrac{A}{x} + \dfrac{B}{x^2} + \dfrac{C}{x^3} + \dfrac{D}{x + 1}$.

d) $\int \tan^3 x \sec^4 x \, dx$

Trigonometric integral,

$\int \tan^3 x \sec^2 x \, (\sec^2 x \, dx)$.

e) $\int x \sec^2 x \, dx$

Integration by parts, $u = x, \, dv = \sec^2 x \, dx$.

f) $\int \dfrac{\csc^2 x}{1 + \cot^2 x} \, dx$

Since $1 + \cot^2 x = \csc^2 x$, this one reduces to

$\int 1 \, dx = x + C$.

B. All the functions you have encountered so far are "elementary," not because they are considered easy, but because it is possible to express them directly as combinations (addition, subtraction, multiplication, division, composition) of polynomial, rational, exponential, logarithmic, trigonometric, inverse trigonometric, hyperbolic, and inverse hyperbolic functions.

The *derivative* of an elementary function *is* elementary.
The *antiderivative* of an elementary function *need not be* elementary.

2) True or False:

$f(x) = e^{x^{x^x}}$ is elementary.

True.

3) True or False:

If $f(x)$ is not elementary, then $\int f(x)\, dx$

is not elementary.

True.

4) True or False:

If $f(x)$ is elementary, then $\int f(x)\, dx$

is elementary.

False.

Using Tables of Integrals

Concepts to Master

Evaluation of integrals by looking in tables

Summary and Focus Questions

Appendix G of your text contains 112 integral formulas. You may need to do some algebra before the integrand fits one of these forms.

What integral formula from the Table of Integrals in the text may be used for each?

1) $\displaystyle\int \frac{dx}{x^2(4 + 5x)}$

          # 50

2) $\displaystyle\int \frac{\sqrt{4x^2 - 9}}{x}\,dx$

          # 41

3) $\displaystyle\int \frac{1}{5x^2 - 3}\,dx$

          # 20

4) $\displaystyle\int \frac{\sqrt{2 + x^2}}{3x}\,dx$

          # 23

Approximate Integration

Concepts to Master

A. Approximation of definite integrals using the Midpoint, Trapezoidal, and Simpson's Rule

B. Maximum error estimation

Summary and Focus Questions

A. Suppose f is an integrable function on a closed interval $[a, b]$ which has been partitioned by $\{x_0, x_1, x_2, \ldots, x_n\}$ into n equal intervals of length $\Delta x = \dfrac{b - a}{n}$.

By the <u>Midpoint Rule</u>, if $\overline{x_i} = \dfrac{x_{i-1} + x_i}{2}$, then

$$\int_a^b f(x)\, dx \approx M_n = \Delta x\, [f(\overline{x_1}) + f(\overline{x_2}) + \ldots + f(\overline{x_n})].$$

M_n is a Riemann sum using the midpoint of each subinterval.

By the <u>Trapezoidal Rule</u>,

$$\int_a^b f(x)\, dx \approx T_n = \frac{\Delta x}{2}[f(x_0) + 2f(x_1) + 2f(x_2) + \ldots + 2f(x_{n-1}) + f(x_n)].$$

This approximation is derived by computing the sums of areas of inscribed trapezoids.

By <u>Simpson's Rule</u>, if n is even,

$$\int_a^b f(x)\, dx \approx S_n = \frac{\Delta x}{3}[f(x_0) + 4f(x_1) + 2f(x_2) + 4f(x_3) + \ldots + 4f(x_{n-1}) + f(x_n)].$$

This approximation is derived by replacing portions of the function $y = f(x)$ with approximating parabolas and summing the areas under these parabolas.

1) Estimate $\int_{-1}^{5} 2^x \, dx$ to three decimal

   places with $n = 6$ subintervals using:

   a) Mipoint Rule

$$\Delta x = \frac{5 - (-1)}{6} = 1.$$

$x_0 = -1, x_1 = 0, x_2 = 1, \ldots, x_6 = 5.$
Thus $\overline{x_1} = -.5, \overline{x_2} = .5, \overline{x_3} = 1.5, \ldots, \overline{x_6} = 4.5.$
$M_n = 1[2^{-.5} + 2^{.5} + 2^{1.5} + \ldots + 2^{4.5}]$
$\quad = 2^{-.5}[1 + 2 + 2^2 + 2^3 + 2^4 + 2^5]$

$$= \frac{63}{\sqrt{2}} \approx 44.547.$$

   b) Trapezoidal Rule

$T_n =$

$\frac{1}{2}[2^{-1} + 2(2^0) + 2(2) + 2(2^2) + 2(2^3) + 2(2^4) + 2^5]$

$$= \frac{1}{2}(94.5) = 47.250.$$

   c) Simpson's Rule

$S_n =$

$\frac{1}{3}[2^{-1} + 4(2^0) + 2(2) + 4(2^2) + 2(2^3) + 4(2^4) + 2^5]$

$$= \frac{1}{3}(136.5) = 45.500.$$

$\left(\text{The actual value of } \int_{-1}^{5} 2^x \, dx \text{ is } \frac{63}{\ln 4} \approx 45.445. \right)$

2) For $\int_1^5 (40 - x)\, dx$ and any n, the

Midpoint approximation will be:
  a) too large
  b) too small
  c) exact
  d) can't tell from the information given.

c) Exact, because $y = 40 - x$ is linear. The area of the midpoint rectangle will equal the area under $y = 40 - x$.

3) For $\int_1^5 (40 - x^2)\, dx$ and any n, the

trapezoidal approximation will be:
  a) too large
  b) too small
  c) exact
  d) can't tell from the information given.

b) Too small. Since the graph of $y = 40 - x^2$ is concave downward, all inscribed trapezoids will have area less than the area under $y = 40 - x^2$.

4) For $\int_1^5 (40 - x^2)\, dx$ and any n, the

Simpson Rule approximation will be:
  a) too large
  b) too small
  c) exact
  d) can't tell from the information given.

c) Exact, because Simpson's Rule replaces the function with portions of approximating parabolas and since $y = 40 - x^2$ is a parabola, the approximation is exact. (If 2 parabolas agree at three points, they must be identical.)

B. The <u>error of estimation</u> is the difference between the actual value of $\int_a^b f(x)\, dx$ and the estimate. Here is a table that describes, under the conditions given, an upper bound on the error of estimation.

Rule	Condition for all $x \in [a, b]$	Maximum Error
Midpoint	$\left\| f''(x) \right\| \leq M$	$\dfrac{M(b-a)^3}{24n^2}$
Trapezoidal	$\left\| f''(x) \right\| \leq M$	$\dfrac{M(b-a)^3}{12n^2}$
Simpson's	$\left\| f^{(4)}(x) \right\| \leq M$	$\dfrac{M(b-a)^5}{180n^4}$

5) Find the maximum error in estimating $\int_{-1}^{3} x^3\, dx$ with $n = 8$ using the Trapezoidal Rule.

The maximum is

$$\frac{M\,(3-(-1))^3}{12(8)^2} = \frac{M}{12}.$$

$f''(x) = 6x$. For $-1 \leq x \leq 3$,
$\left| f''(x) \right| = \left| 6x \right| \leq 6 \cdot 3 = 18$.
Thus the maximum error of estimation is

$$\frac{M}{12} = \frac{18}{12} = 1.5.$$

6) Find the maximum error in estimating

$$\int_1^3 \sin 2x \, dx \text{ with } n = 4 \text{ using}$$

Simpson's Rule.

$$\frac{M(3-1)^5}{180(4)^4} = \frac{M}{1440}.$$

For $f(x) = \sin 2x$, $f^{(4)}(x) = 16 \sin 2x$.

For $1 \le x \le 3$, $|\sin 2x| \le 1$, so $|f^{(4)}(x)| \le 16$.

Thus the maximum error is

$$\frac{M}{1440} = \frac{16}{1440} = \frac{1}{90}.$$

Improper Integrals

Concepts to Master

A. Definition and evaluation of improper integrals; Type I; Type II; Convergent and Divergent

B. Comparison Test for Integrals

Summary and Focus Questions

A. Improper integrals are defined as follows:

<u>Definition</u>

a) $\displaystyle\int_a^\infty f(x)\, dx = \lim_{t \to \infty}\int_a^t f(x)\, dx$

b) $\displaystyle\int_{-\infty}^b f(x)\, dx = \lim_{t \to -\infty}\int_t^b f(x)\, dx$

c) $\displaystyle\int_{-\infty}^\infty f(x)\, dx = \int_{-\infty}^a f(x)\, dx + \int_a^\infty f(x)\, dx$

d) $\displaystyle\int_a^b f(x)\, dx = \lim_{t \to a^+}\int_t^b f(x)\, dx$

e) $\displaystyle\int_a^b f(x)\, dx = \lim_{t \to b^-}\int_a^t f(x)\, dx$

f) $\displaystyle\int_a^b f(x)\, dx = \int_a^c f(x)\, dx + \int_c^b f(x)\, dx$

<u>Necessary Condition</u>

$\displaystyle\int_a^t f(x)\, dx$ exists for all $t \geq a$.

$\displaystyle\int_t^b f(x)\, dx$ exists for all $t \leq b$.

Both integrals on the right exist.

$\displaystyle\int_t^b f(x)\, dx$ exists for $t \in (a, b]$.

$\displaystyle\int_a^t f(x)\, dx$ exists for all $t \in [a, b)$.

Both improper integrals on the right exist.

Section 7.9: Improper Integrals

If an improper integral exists, it is said to <u>converge</u>; otherwise it diverges. Improper integrals are evaluated as limits according to their definitions.

1) Give a definition of each:

   a) $\int_0^\infty \sin 2x \, dx$

   $$\lim_{t \to \infty} \int_0^t \sin 2x \, dx$$

   b) $\int_0^1 \frac{1}{\sqrt{x}} dx$

   $$\lim_{t \to 0^+} \int_t^1 \frac{1}{\sqrt{x}} dx$$

   c) $\int_{-\infty}^\infty e^{-x^2} dx$

   $$\int_{-\infty}^0 e^{-x^2} dx + \int_0^\infty e^{-x^2} dx$$

   d) $\int_{-2}^2 \frac{1}{(x-1)^2} dx$

   $$\int_{-2}^1 \frac{1}{(x-1)^2} dx + \int_1^2 \frac{1}{(x-1)^2} dx$$

2) Evaluate $\int_0^{\pi/2} \sec^2 x \, dx$.

   $$\int_0^{\pi/2} \sec^2 x \, dx = \lim_{t \to \frac{\pi}{2}^-} \int_0^t \sec^2 x \, dx$$
   $$= \lim_{t \to \frac{\pi}{2}^-} \tan x \Big|_0^t$$
   $$= \lim_{t \to \frac{\pi}{2}^-} \tan t, \text{ which does not exist}$$

   (the limit is ∞). This integral diverges.

3) Evaluate $\displaystyle\int_1^\infty x\,e^{-x^2}dx$.

$$\int_1^\infty x\,e^{-x^2}dx = \lim_{t\to\infty}\int_1^t x\,e^{-x^2}dx$$

$$(u=-x^2,\ du=-2x\,dx)$$

$$= \lim_{t\to\infty}\ -\frac{1}{2}\,e^{-x^2}\Big|_1^t$$

$$= \lim_{t\to\infty}\ -\frac{1}{2}(e^{-x^2}-e^1)$$

$$= -\frac{1}{2}\,(0-e)=\frac{e}{2}.$$

4) a) Find the area of the shaded region.

$$\int_1^\infty \frac{1}{x}dx = \lim_{t\to\infty}\int_1^t\frac{1}{x}dx$$

$$= \lim_{t\to\infty}\ \ln t\ \Big|_1^t = \lim_{t\to\infty}\ \ln t = \infty.$$

The improper integral does not exist; the area is infinite.

b) Evaluate $\displaystyle\int_1^\infty \frac{\pi}{x^2}dx$.

$$\int_1^\infty \frac{\pi}{x^2}dx = \lim_{t\to\infty}\int_1^t\frac{\pi}{x^2}dx$$

$$= \lim_{t\to\infty}\frac{-\pi}{x}\Big|_1^t = \lim_{t\to\infty}\frac{-\pi}{x}+\frac{\pi}{1}=\pi.$$

c) (for fun) The region in part a) has an infinite area. How can it be painted with a finite amount of paint?

Step 1. Take the curve $y = \dfrac{1}{x}$ and revolve it about the x-axis.

Step 2. The volume of the resulting solid of revolution is given by $\displaystyle\int_1^\infty \dfrac{\pi}{x^2}\,dx$, which by part b) is π.

Step 3. Fill the solid of revolution with π gallons of paint.

Step 4. Dip the region in part a) in the solid of revolution. The region is then painted with at most π gallons of paint!

B. Sometimes improper integrals can be shown to converge (although not necessarily to a determined value) without attempting to evalutate them. One such method is the Comparison Test for Integrals:

Suppose $f(x)$ and $g(x)$ are continuous and $f(x) \geq g(x) \geq 0$ for $x \geq a$.

If $\displaystyle\int_a^\infty f(x)\,dx$ converges then $\displaystyle\int_a^\infty g(x)\,dx$ converges.

Similar statements can be made for the other types of improper integrals.

5) True or False:

If $f(x) \geq g(x) \geq 0$ for $x \geq a$ and f and g are continuous, then if $\displaystyle\int_a^b g(x)\,dx$ diverges, then $\displaystyle\int_a^\infty f(x)\,dx$ diverges.

True. This statement is just the contrapositive of the Comparison Test for Integrals.

6) Show that $\displaystyle\int_1^\infty \frac{|\sin x|}{x^2}\, dx$ converges.

For $1 \le x \le \infty$, $0 \le |\sin x| \le 1$.

Thus $0 \le \dfrac{|\sin x|}{x^2} \le \dfrac{1}{x^2}$.

$\displaystyle\int_1^\infty \frac{1}{x^2}\, dx$ converges (to 1). Thus $\displaystyle\int_1^\infty \frac{|\sin x|}{x^2}\, dx$ converges, (but we don't know to what value).

Further Applications of Integration

"THAT WRAPS IT UP — THE MASS OF THE UNIVERSE."

Differential Equations

Concepts to Master

Order, degree, general and particular solution to a differential equation;
Separable equations

Summary and Focus Questions

A <u>differential equation</u> is an equation involving x, y, y', y'', ... $y^{(n)}$. The n in the highest $y^{(n)}$ is the <u>order</u> of the equation and the <u>degree</u> is the exponent that $y^{(n)}$ has. For example, $2(y''')^4 + 5xy' + 7x = 0$ has degree 4 and order 3.

A <u>particular solution</u> to a differential equation is a function $y = f(x)$ that satisfies the equation. A <u>general solution</u> is an expression with arbitrary constants to represent all particular solutions. <u>Initial boundary conditions</u> are values used to determine a particular solution from the general.

A <u>separable differential equation</u> is an equation that can be written as $\dfrac{dy}{dx} = \dfrac{f(x)}{g(y)}$.

It is solved by rewriting as $g(y)dy = f(x)dx$, integrating to get

$\int g(y)dy = \int f(x)dx + C$, and solving for y.

1) The equation $x^2(y'')^3 + 4xy' - 2y + x = 0$
   has degree _____ and order _____.

| 3 2

2) Is $y' = xy + x$ separable?

$$\underline{\text{Yes.}} \ \frac{dy}{dx} = x(y + 1) = \frac{x}{(y + 1)^{-1}}.$$

3) Solve the equation $y' = \dfrac{\sin x}{\cos y}$.

The equation is separable

$$\frac{dy}{dx} = \frac{\sin x}{\cos y}$$

$$\cos y \, dy = \sin x \, dx$$

$$\int \cos y \, dy = \int \sin x \, dx$$

$$\sin y = -\cos x + C$$
$$y = \sin^{-1}(-\cos x + C).$$

4) Find the particular solution for the initial condition $y(0) = 2$ for the equation $e^{x^2} yy' + x = 0.$

First separate: $e^{x^2} y \dfrac{dy}{dx} = -x$

$$y \frac{dy}{dx} = -x e^{-x^2}$$

Integrate: $\int y \, dy = \int -x e^{-x^2} \, dx$

$$\frac{y^2}{2} = \frac{1}{2} e^{-x^2} + C$$

$$y^2 = e^{-x^2} + C$$

$$y = (e^{-x^2} + C)^{\frac{1}{2}} = \sqrt{e^{-x^2} + C}$$

Find the particular solution:

At $x = 0$, $y = 2$ so $\sqrt{e^{-0} + C} = 2$.

Thus $1 + C = 4$, $C = 3$.

The particular solution is $y = \sqrt{e^{-x^2} + 3}$.

Arc Length

Concepts to Master

A. Length of a curve
B. Arc length function and its derivative

Summary and Focus Questions

A. If f' is continuous on $[a, b]$, the <u>length of the curve</u> $y = f(x)$, $a \le x \le b$ (arc length) is

$$L = \int_a^b \sqrt{1 + (f'(x))^2}\, dx.$$

For curves described as $x = g(y)$ for $c \le y \le d$, use $\int_c^d \sqrt{1 + (g'(y))^2}\, dy$.

1) Find a definite integral for the length of

$$y = \frac{1}{x}, \ 1 \le x \le 3.$$

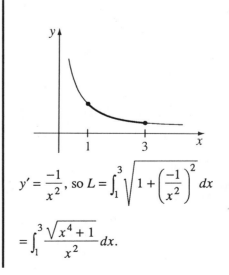

$y' = \dfrac{-1}{x^2}$, so $L = \displaystyle\int_1^3 \sqrt{1 + \left(\dfrac{-1}{x^2}\right)^2}\, dx$

$= \displaystyle\int_1^3 \dfrac{\sqrt{x^4 + 1}}{x^2}\, dx.$

2) What is the length of the arc
$x = 1 + y^{3/2}$, $0 \le y \le 4$?

$$x' = \frac{3}{2} y^{1/2}$$

$$L = \int_0^4 \sqrt{1 + \left(\frac{3}{2} y^{1/2}\right)^2} \, dy = \int_0^4 \sqrt{1 + \frac{9}{4} y} \, dy$$

$$\left(u = 1 + \frac{9}{4}y, \, du = \frac{9}{4} dy\right)$$

$$= \frac{4}{9} \int \sqrt{u} \, du = \frac{4}{9} \frac{u^{3/2}}{3/2}$$

$$= \frac{8}{27} \left(1 + \frac{9}{4}y\right)^{3/2} \Big|_0^4$$

$$= \frac{8}{27}\left(10\sqrt{10} - 1\right).$$

3) Write an expression involving definite integrals for the length of the curve

$y = \sqrt[3]{x}$ from $x = -1$ to $x = 8$.

Since a vertical tangent exists at $x = 0$, we could divide the curve at $(0, 0)$ into arc I and arc II. The curve length is the sum of the lengths of these two arcs. Alternatively, treat x as a function of y. From

$y = \sqrt[3]{x}$, $x = y^3$ and $x' = 3y^2$. The arc length is

$$\int_{-1}^{2} \sqrt{1 + (3y^2)^2} \, dy = \int_{-1}^{2} \sqrt{1 + 9y^4} \, dy.$$

B. For a smooth function $y = f(x)$, $a \le x \le b$, let $s(x)$ be the distance along the curve from $(a, f(a))$ to $(x, f(x))$.

$$s(x) = \int_a^x \sqrt{1 + [f'(t)]^2} \; dt$$

The differential $ds = \sqrt{1 + \left(\dfrac{dy}{dx}\right)^2} \; dx$ may be written $(ds)^2 = (dy)^2 + (dx)^2$.

4) Find the arc length function for
   $y^2 = x^3$, $0 \le x \le 1$.

$y^2 = x^3$, $y = x^{3/2}$

$\dfrac{dy}{dx} = 3x^{1/2}$

$s(x) = \int_0^x \sqrt{1 + (3t^{1/2})^2} \; dt = \int_0^x \sqrt{1 + 9t} \; dt$

$(u = 1 + 9t, \; du = 9 \; dt)$

$= \dfrac{1}{9} \dfrac{(1 + 9t)^{3/2}}{3/2}$

$= \dfrac{2}{27}(1 + 9t)^{3/2} \Big|_0^x$

$= \dfrac{2}{27}(1 + 9x)^{3/2} - \dfrac{2}{27}.$

Area of a Surface of Revolution

Concepts to Master

Computation of surface area of a solid of revolution

Summary and Focus Questions

If $f(x) \geq 0$ and $f'(x)$ is continuous on $[a, b]$, the <u>surface area</u> of the solid of revolution obtained by rotating $f(x)$ about the x-axis is

$$S = \lim_{\|P\| \to 0} \sum_{i=1}^{n} 2 \pi f(x_i^*)\sqrt{1 + (f'(x_i^*))^2} \, \Delta x_i$$

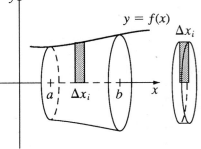

$$= \int_a^b 2 \pi f(x)\sqrt{1 + (f'(x))^2} \, dx.$$

For curves described by $x = g(y)$, $c \leq y \leq d$,

$$S = \int_c^d 2 \pi y \sqrt{1 + \left(\frac{dx}{dy}\right)^2} \, dy.$$

Using shorter notation that represents both of the above cases of rotation about the x-axis:

$$S = \int 2 \pi y \, ds, \text{ where } ds = \sqrt{1 + \left(\frac{dy}{dx}\right)^2} \, dx \text{ or } ds = \sqrt{1 + \left(\frac{dx}{dy}\right)^2} \, dy$$

For rotation about the y-axis, $S = \int 2 \pi x \, ds$.

Section 8.3: Area of a Surface of Revolution

1) Find the surface area obtained by revolving $y = \sin 2x$, $0 \le x \le \pi/2$, about the x-axis.

$$S = \int_0^{\pi/2} 2\pi \sin 2x \sqrt{1 + (2 \cos 2x)^2}\, dx$$

$$\left(\begin{array}{l} u = 2 \cos 2x,\ du = -4 \sin 2x \\ \text{At } x = 0,\ u = 2;\ x = \pi/2,\ u = -2 \end{array} \right)$$

$$= \frac{-\pi}{2} \int_2^{-2} \sqrt{1 + u^2}\, du = \frac{\pi}{2} \int_{-2}^{2} \sqrt{1 + u^2}\, du$$

(Table of Integrals, #21)

$$= \frac{u}{2} \sqrt{1 + u^2} + \frac{1}{2} \ln \left| u + \sqrt{1 + u^2} \right| \ \Big|_{-2}^{2}$$

$$= \frac{2}{2}\sqrt{5} + \frac{1}{2}\ln\left| 2 + \sqrt{5}\, \right|$$

$$\qquad - \left(\frac{-2}{2}\sqrt{5} + \frac{1}{2}\ln\left| 2 + \sqrt{5}\, \right| \right)$$

$$= 2\sqrt{5} + \ln\left(2 + \sqrt{5}\, \right).$$

2) Set up two equivalent definite integrals for the surface area obtained by revolving $y = x^2$, $0 \le x \le 2$ about the y-axis.

In both solutions $S = \int 2\pi x\, ds$.

I. $ds = \sqrt{1 + \left(\dfrac{dy}{dx} \right)^2}\, dx$

$$S = \int_0^2 2\pi x \sqrt{1 + (2x)^2}\, dx = \int_0^2 2\pi x \sqrt{1 + 4x^2}\, dx$$

II. $ds = \sqrt{1 + \dfrac{dx}{dy}}\ dy$

Since $y = x^2$, $0 \le x \le 2$,
$x = \sqrt{y}$, $0 \le y \le 4$.

$$S = \int_0^4 2\pi \sqrt{y}\ \sqrt{1 + \left(\dfrac{1}{2\sqrt{y}}\right)^2}\ dy$$

$$= \int_0^4 2\pi \sqrt{y}\ \sqrt{\dfrac{1 + 4y}{4y}}\ dy$$

$$= \int_0^4 \pi\sqrt{1 + 4y}\ dy.$$

This solution is equivalent to the integral in part I since $y = x^2$, $dy = 2x\ dx$, and for $0 \le x \le 2$, we have $0 \le y \le 4$.

Moments and Centers of Mass

Concepts to Master

A.   Moments about x-axis and y-axis; Centroid (center of mass)
B.   Theorem of Pappus

Summary and Focus Questions

A. A lamina is a thin flat sheet of material
    determined by the region $\mathfrak{R}$, in the figure
    whose density is uniformly ρ.

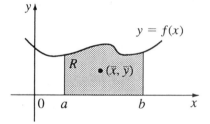

Let $A = \displaystyle\int_a^b f(x)\, dx$ be the area of $\mathfrak{R}$.

The <u>centroid</u> (or <u>center of mass</u>) of $\mathfrak{R}$ is

the point $(\bar{x}, \bar{y})$ where

$$\bar{x} = \frac{1}{A} \int_a^b x\, f(x)\, dx, \quad \bar{y} = \frac{1}{A} \int_a^b \frac{1}{2}\, [f(x)]^2\, dx.$$

The <u>moment of $\mathfrak{R}$ about the x-axis</u> is $M_x = \rho \displaystyle\int_a^b \frac{1}{2}[f(x)]^2\, dx$ and the moment

about the y-axis is $M_y = \rho \displaystyle\int_a^b x\, f(x)\, dx$.

1) Find M_x, M_y, and the center of mass of the region bounded by $y = 4 - x^2$, $x = 0$, $y = 0$.

The area of the region is

$$A = \int_0^2 (4 - x^2)\, dx = 4x - \frac{x^3}{3}\ \bigg|_0^2$$

$$= \frac{16}{3}, \text{ so } \frac{1}{A} = \frac{3}{16}.$$

For M_y and $\bar{x}$:

$$\int_a^b x f(x)\, dx = \int_0^2 x(4 - x^2)\, dx$$

$$= \int_0^2 (4x - x^3)\, dx = \left(2x^2 - \frac{x^4}{4}\right)\bigg|_0^2 = 4.$$

Therefore $M_y = 4\rho$ and $\bar{x} = \left(\frac{3}{16}\right)(4) = \frac{3}{4}$.

For M_x and $\bar{y}$:

$$\int_a^b \frac{1}{2}[f(x)]^2\, dx = \frac{1}{2}\int_0^2 (4 - x^2)^2\, dx$$

$$= \frac{1}{2}\int_0^2 (16 - 8x^2 + x^4)\, dx$$

$$= \frac{1}{2}\left(16x - \frac{8}{3}x^3 + \frac{x^5}{5}\right)\bigg|_0^2 = \frac{128}{15}.$$

Therefore $M_x = \frac{128}{15}\rho$ and $\bar{y} = \frac{3}{16}\frac{128}{15} = \frac{8}{5}$.

B. The <u>Theorem of Pappus</u> is: If a region $\mathfrak{R}$ is revolved about a line L that does not intersect $\mathfrak{R}$, the resulting solid has volume $V = 2\pi dA$ where A is the area of $\mathfrak{R}$ and d is the perpendicular distance between the centroid of $\mathfrak{R}$ and the line L. (Notice that $2\pi d$ is the distance traveled by the centroid.)

2) Find the volume of the solid of revolution obtained by rotating the region in question 1 about

    a) the x-axis.

$\bar{y} = \dfrac{8}{5}$ is distance from the center of mass to the axis of revolution (the x-axis). The area is $A = \dfrac{16}{3}$.

Thus $V = 2\pi \left(\dfrac{8}{3}\right)\left(\dfrac{16}{3}\right) = \dfrac{256\pi}{3}$.

    b) the y-axis.

$\bar{x} = \dfrac{3}{4}$, so

$V = 2\pi \left(\dfrac{3}{4}\right)\left(\dfrac{16}{3}\right) = 8\pi$.

Hydrostatic Pressure and Force

Concepts to Master

Calculation of force on a vertical plate due to hydrostatic pressure; Calculation of pressure at a given depth.

Summary and Focus Questions

The pressure on a plate suspended horizontally in a liquid with density ρ at a depth d is

$$P = \rho g d,$$

where g is the gravitational constant. P is in units of newtons per meter^2 (= pascals) or in pounds per ft^3.

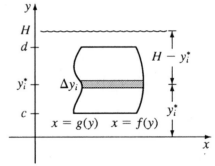

Suppose a plate is suspended vertically in a liquid with mass density ρ at depth H and the surface of the plate can be described as bounded by $x = f(y)$, $x = g(y)$, $y = c$, $y = d$ with $f(y) \geq g(y)$ for all $y \in [c, d]$. Then the force due to liquid pressure on a section of the plate is approximately

(density)(gravitational constant)(depth)(area of section)

$$= \rho g (H - y_i^*)[f(y_i^*) - g(y_i^*)]\Delta y_i.$$

Thus the total force on the surface is

$$\int_c^d \rho g (H - y)(f(y) - k(y))dy.$$

1) A circular plate of radius 3 cm is suspended horizontally at a depth of 12 m in an oil having density 1500 kg/m³. What is the pressure on the plate?

$P = \rho\, g\, d$
$= (1500 \text{ kg/m}^3)(9.8 \text{ m/s}^2)(12 \text{ m})$
$= 176{,}400 \text{ Pa (pascals)} = 176.4 \text{ kPa.}$

2) A flat isoceles triangle is suspended in water as in the figure. Water density is 1000 kg/m³. Set up a definite integral for the hydrostatic force.

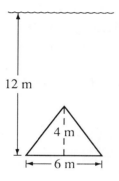

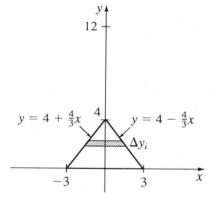

Setting up coordinates as shown the sides of the triangle are

$y = 4 - \dfrac{4}{3}x$, so $x = \dfrac{-3}{4}y + 3$, and

$y = 4 + \dfrac{4}{3}x$, so $x = \dfrac{3}{4}y - 3$

The force on the triangle is

$$\int_0^4 1000(9.8)(12 - y)\left[\left(-\frac{3}{4}y + 3\right) - \left(\frac{3}{4}y - 3\right)\right] dy$$

$$= \int_0^4 9800(12 - y)\left(6 - \frac{3}{2}y\right) dy.$$

Applications to Economics and Biology

Concepts to Master

Using a definite integral to determine the total value of a quantity

Summary and Focus Questions

Applications of definite integrals in this section and elsewhere all involve calculating the total amount of a quantity (that can be represented by a continuous $y = f(x)$) within a certain range ($a \leq x \leq b$). The idea of summing the values (Riemann sums) is extended to the continuous case $\left(\lim_{\|P\| \to 0} \right)$ to give the definite integral ($\int_a^b f(x)\, dx$). Here are two examples from economics:

1. For a demand function $p(x)$, the price necessary to sell x items, let X be the amount actually available and $P = p(X)$.
   The <u>consumers' surplus</u> is

   $$\int_0^X [p(x) - P]dx,$$

   the total amount of money that could be saved by consumers if they buy when the price is P dollars.

2. If $f(t)$ is the amount of money that is to be received at time $t \in [a, b]$ and interest is compounded continuously at $r\%$, the <u>present value</u> of $f(t)$ is

   $$\int_a^b e^{-rt} f(t)\, dt,$$

   the total amount of money that must be invested now so that the amounts collected at times $t(a \leq t \leq b)$ will be $f(t)$ dollars.

1) Find a definite integral for the consumer surplus determined by 1000 units available and a demand function of $p(x) = 85 - .03x$

At $X = 1000$, $P = 85 - .03(1000) = 75$.

The consumer surplus is

$$\int_0^{1000} [85 - .03x - 75] \, dx$$

$$= \int_0^{1000} (10 - .03x) \, dx$$

2) A family sets up a trust fund for a 10-year-old daughter so that between the ages of 10 and 21 she will receive $1000 the first year and double the amount each following year. How much needs to be invested now (what is the present value) if interest is compounded continuously at 7%?

$f(t) = 1000 \, 2^t$, $t \in [0, 11]$.

The present value is $\int_0^{11} e^{-.07t}(1000 \, 2^t) \, dt$.

9

Parametric Equations and Polar Coordinates

Curves Defined by Parametric Equations

Concepts to Master

Parametric equations; Graphs of curves defined by parametric equations; Elimination of the parameter

Summary and Focus Questions

A set of <u>parametric equations</u> has the form
$$x = f(t), \, y = g(t)$$
where f and g are functions of a third variable t called a <u>parameter</u>. To each t there corresponds an x value and a y value, hence a point in the xy plane. The collection of all such points is a <u>curve</u>. The same curve may be described by several different pairs of equations.

Some pairs of parametric equations can be reduced algebraically to a single equation not involving t; this is called <u>eliminating the parameter</u>. The method used depends greatly on the nature of the functions involved. Often one equation can be solved for t and the result substituted for t in the other.

<u>Example</u>: Eliminate the parameter in $x = 2 + 3t, \, y = t^2 - 2t + 1$.

Solve $x = 2 + 3t$ for t to get $t = \dfrac{1}{3}(x - 2)$. Substitute this in the other equation:

$$y = \left[\frac{1}{3}(x - 2)\right]^2 - 2\left[\frac{1}{3}(x - 2)\right] + 1 \quad or \quad y = \frac{1}{9}x^2 - \frac{10}{9}x + \frac{25}{9}.$$

If trigonometric functions are involved, frequently a trigonometric identity can be employed:

Example: Eliminate the parameter in $x = \sin t$, $y = \cot^2 t$.

$$y = \cot^2 t = \frac{\cos^2 t}{\sin^2 t} = \frac{1 - \sin^2 t}{\sin^2 t}. \text{ Thus } y = \frac{1 - x^2}{x^2}.$$

We note from $x = \sin t$, $-1 \le x \le 1$ and $x \ne 0$ (since when $\sin t = 0$, $\cot t$ is undefined).

1) Sketch a graph of the curve given by
   $x = \sqrt{t}$, $y = t + 2$.

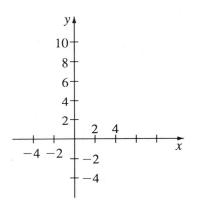

Compute some values and plot points:

t	0	1	4	9
x	0	1	2	3
y	2	3	6	11

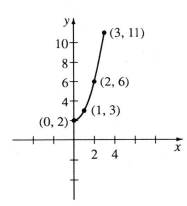

We see this is half a parabola since $x = \sqrt{t}$
implies $x^2 = t$, thus $y = x^2 + 2$, $x \ge 0$.

2) Eliminate the parameter in each
   a) $x = e^t$, $y = t^2$

Solve for t in terms of y:
$y = t^2$, $\sqrt{y} = t$.
Thus $x = e^{\sqrt{y}}$, $x > 0$.
A different solution may be obtained by solving
for t in terms of x:
$x = e^t$, $t = \ln x$, $y = (\ln x)^2$, $x > 0$.

b) $x = 1 + \cos t, y = \sin^2 t$

$x = 1 + \cos t, x - 1 = \cos t,$
$\cos^2 t = (x - 1)^2.$
Since $\sin^2 t = y$, and $\cos^2 t + \sin^2 t = 1$ we have
$(x - 1)^2 + y = 1.$
Thus $y = 1 - (x - 1)^2.$
Since $x = 1 + \cos t, 0 \leq x \leq 2.$

c) $x = 2 \sec t, y = 3 \tan t$

$\sec t = \dfrac{x}{2}$ and $\tan t = \dfrac{y}{3}.$

Hence the identity $\tan^2 t + 1 = \sec^2 t$

becomes $\left(\dfrac{y}{3}\right)^2 + 1 = \left(\dfrac{x}{2}\right)^2$

or $\dfrac{x^2}{4} - \dfrac{y^2}{9} = 1.$

Tangents and Areas

Concepts to Master

A. First and second derivative of a function defined by a pair of parametric equations
B. Area under a curve defined by a pair of parametric equations

Summary and Focus Questions

A. The slope of the line tangent to a curve described by $x = f(t)$, $y = g(t)$ is the first derivative of y with respect to x and is given by

$$\frac{dy}{dx} = \frac{\dfrac{dy}{dt}}{\dfrac{dx}{dt}} \quad \text{if } \frac{dx}{dt} \neq 0.$$

In the case $\dfrac{dx}{dt} = 0$ but $\dfrac{dy}{dt} \neq 0$, the curve will have a vertical tangent line.

The second derivative of y with respect to x is:

$$\frac{d^2y}{dx^2} = \frac{\dfrac{d}{dt}\left(\dfrac{dy}{dx}\right)}{\dfrac{dx}{dt}}.$$

1) For $x = t^3$, $y = t^2 - 2t$

  a) Find $\dfrac{dy}{dx}$ and $\dfrac{d^2y}{dx^2}$.

$\dfrac{dx}{dt} = 3t^2$ and $\dfrac{dy}{dt} = 2t - 2$, so $\dfrac{dy}{dx} = \dfrac{2t - 2}{3t^2}$.

$$\frac{d}{dt}\left(\frac{dy}{dx}\right) = \frac{3t^2(2) - (2t - 2)(6t)}{(3t^2)^2} = \frac{4 - 2t}{3t^3}.$$

$$\text{Thus } \frac{d^2y}{dx^2} = \frac{\dfrac{4 - 2t}{3t^3}}{3t^2} = \frac{4 - 2t}{9t^5}.$$

b) Find the horizontal and vertical tangents for the curve.

$\dfrac{dy}{dx} = 0$ at $t = 1$ $(x = 1, y = -1)$.

A horizontal tangent is at $(-1, 1)$.

$\dfrac{dy}{dx}$ does not exist at $t = 0$ $(x = 0, y = 0)$.

A vertical tangent is at $(0, 0)$.

c) Discuss the concavity of the curve.

$\dfrac{d^2y}{dx^2} > 0$ for $0 < t < 1$ $(0 < x < 1)$ and negative for

$t < 0$ $(x < 0)$ and $t > 1$ $(x > 1)$.
The curve is concave upward for
$x \in (0, 1)$ and concave downward for
$x \in (-\infty, 0)$ and $x \in (1, \infty)$.

d) Sketch the curve.

t	-2	-1	0	1	2	3
x	-8	-1	0	1	8	27
y	8	3	0	-1	0	3

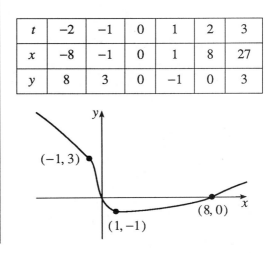

$(-1, 3)$

$(8, 0)$

$(1, -1)$

2) Find the equation of the tangent line to
the curve $x = 4t^2 + 2t + 1$, $y = 7t + 2t^2$
at the point corresponding to $t = -1$.

At $t = -1$, $x = 3$ and $y = -5$.

$\dfrac{dx}{dt} = 8t + 2 = -6$ at $t = -1$.

$\dfrac{dy}{dt} = 7 + 4t = -3$ at $t = -1$.

Thus $\dfrac{dy}{dx} = \dfrac{-3}{-6} = \dfrac{1}{2}$. The tangent line is

$y + 5 = \dfrac{1}{2}(x - 3)$.

B. Suppose the parametric equations $x = f(t)$, $y = g(t)$, $t \in [t_1, t_2]$ define an integrable
function $y = F(x) \geq 0$ over the interval $[a, b]$, where $a = f(t_1)$, $b = f(t_2)$, and

$f'(t) \geq 0$. The area under $y = F(x)$ is $\displaystyle\int_a^b y \, dx = \int_{t_1}^{t_2} g(t) f'(t) \, dt$.

3) Find a definite integral for the area under
the curve $x = t^2 + 1$, $y = e^t$, $0 \leq t \leq 1$.

The area is $\displaystyle\int_0^1 e^t \, (2t) \, dt$

$= \left(\text{by parts:} \quad \begin{matrix} u = 2t & dv = e^t dt \\ du = 2 \, dt & v = e^t \end{matrix} \right)$

$= 2t \, e^t - \displaystyle\int e^t \, 2 \, dt$

$= 2t \, e^t - 2e^t \Big|_0^1$

$= (2e - 2e) - (0 - 2) = 2.$

Arc Length and Surface Area

Concepts to Master

A. Length of a curve defined parametrically
B. Area of a surface of revolution defined parametrically

Summary and Focus Questions

A. The length of a curve defined by $x = f(t)$, $y = g(t)$, $t \in [\alpha, \beta]$ with f', g' continuous and the curve traversed only once as t increases from α to β is

$$\int_\alpha^\beta \sqrt{\left(\frac{dx}{dt}\right)^2 + \left(\frac{dy}{dt}\right)^2}\, dt$$

Since $(ds)^2 = (dx)^2 + (dy)^2$ (see Section 8.2) this is $\int ds$.

1) Write a definite integral for the arc length of each:

a) $x = \ln t$, $y = t^2$, for $1 \leq t \leq 2$.

$$\int_1^2 \sqrt{\left(\frac{1}{t}\right)^2 + (2t)^2}\, dt$$

b) an ellipse given by

$$\frac{x^2}{a^2} + \frac{y^2}{b^2} = 1, \, a, b > 0.$$

The ellipse may be defined by
$x = a \cos t$, $y = b \sin t$, $t \in [0, 2\pi]$.
The length is

$$\left. \int_0^{2\pi} \sqrt{(-a \sin t)^2 + (b \cos t)^2} \; dt \right.$$

$$= \int_0^{2\pi} \sqrt{a^2 \sin^2 t + b^2 \cos^2 t} \; dt$$

B. If a curve given by $x = f(t)$, $y = g(t)$, $t \in [\alpha, \beta]$, with f', g' continuous and $g(t) \geq 0$ is rotated about the x-axis, the area of the surface of revolution is

$$S = \int_\alpha^\beta 2\pi y \sqrt{\left(\frac{dx}{dt}\right)^2 + \left(\frac{dy}{dt}\right)^2} \; dt$$

Using the ds notation this is $S = \int 2\pi y \; ds$, the same formula as in Section 8.3.

2) Find a definite integral for the area of the surface of revolution about the x-axis for each curve:

a) $x = 2t + 1$, $y = t^3$, $t \in [0, 2]$

$$S = \int_0^2 2\pi(t^3)\sqrt{(2)^2 + (3t^2)^2} \; dt$$

$$= \int_0^2 2\pi t^3 \sqrt{4 + 9t^4} \; dt.$$

b) a "football" obtained from rotating the

ellipse $\dfrac{x^2}{a^2} + \dfrac{y^2}{b^2} = 1$, $y \geq 0$.

The top half of the ellipse is $x = a \cos t$,
$y = b \sin t$, $t \in [0, \pi]$. $ds = \sqrt{(dx)^2 + (dy)^2}$
$= \sqrt{(-a \sin t)^2 + (b \cos t)^2}$
$= \sqrt{a^2 \sin^2 t + b^2 \cos^2 t}$.

Thus the area S

$$= \int_0^\pi 2\pi(b \sin t)\sqrt{a^2 \sin^2 t + b^2 \cos^2 t} \; dt.$$

Polar Coordinates

Concepts to Master

A. Plotting points in polar coordinates; Conversion to and from rectangular coordinates

B. Graphs of polar equations; Tangents to polar curves

Summary and Focus Questions

A. To construct a <u>polar coordinate system</u> fix a point called the <u>pole</u> and a ray from the pole called the <u>polar axis</u>. Then a point P has polar coordinates (r, θ) if

$|r|$ = the distance from the pole to P, and

θ = the measure of a directed angle with initial side the polar axis and terminal side the line through the pole and P.

To plot a point P with polar coordinates $P(r, \theta)$:

if $r = 0$, P is the pole (for any value of θ).

if $r > 0$, rotate the polar axis by the angle θ (counterclockwise for $\theta > 0$, clockwise for $\theta < 0$) and locate P on this ray at a distance r units from the pole.

if $r < 0$, rotate the polar axis by the angle θ and then reflect about the pole. P is located on this reflected ray at a distance $-r$ units from the pole.

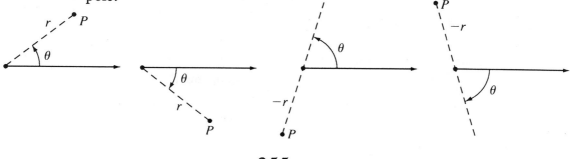

If a rectangular coordinate system
is placed upon the polar coordinate
system as in the figure on the right,
then

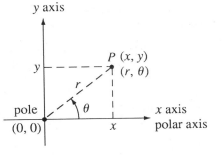

to change from polar to rectangular:
$$x = r \cos \theta$$
$$y = r \sin \theta$$
to change from rectangular to polar:

θ is a solution to $\tan \theta = \dfrac{y}{x}$ (for $x \neq 0$).

r is a solution to $r^2 = x^2 + y^2$ where $r > 0$ if the terminal side of
the angle θ is in the same quadrant as P; if not, then $r \leq 0$.

1) Plot these points in polar coordinates.

A: $(4, \frac{\pi}{3})$,   B: $(-3, \frac{\pi}{4})$
C: $(0, 3\pi)$,   D: $(2, -\frac{\pi}{4})$

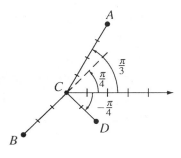

2) Find 2 other polar coordinates for $(8, \frac{\pi}{3})$.

There are infinitely many answers, including

$$\left(8, \frac{7\pi}{3}\right), \left(-8, \frac{4\pi}{3}\right), \left(8, \frac{-5\pi}{3}\right), \dots$$

3) Sometimes, Always, or Never:

    a) The polar coordinates of a point are
       unique.

<u>Never.</u>

    b) $(r, \theta) = (r, \theta + 2\pi)$.

<u>Always.</u>

    c) $(r, \theta) = (-r, \theta + \pi)$.

<u>Always.</u>

    d) $(r, \theta) = (-r, \theta)$.

<u>Sometimes.</u> (True when $r = 0$).

4) Find polar coordinates for the point P with
    rectangular coordinates $P: \left(-3, -3\sqrt{3}\right)$.

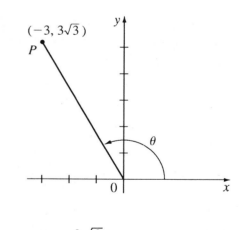

$\tan \theta = \dfrac{y}{x} = \dfrac{3\sqrt{3}}{-3} = -\sqrt{3}$.

A solution is $\theta = \dfrac{2\pi}{3}$.

$r^2 = (-3)^2 + \left(3\sqrt{3}\right)^2 = 9 + 27 = 36$.

Because the terminal side of $\theta = \dfrac{2\pi}{3}$ lies in the same quadrant as P, $r > 0$.

Therefore, P: $(6, \frac{2\pi}{3})$.

Note: $\theta = \frac{-\pi}{6}$ is another solution to $\tan \theta = -\sqrt{3}$,

which results in $r = -6$ and coordinates $(-6, -\frac{\pi}{6})$.

5) Find the rectangular coordinates for
   the point with polar coordinates

   $(-4, \frac{5\pi}{6})$.

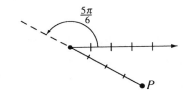

$x = -4 \cos \frac{5\pi}{6} = (-4)\left(-\frac{\sqrt{3}}{2}\right) = 2\sqrt{3}$.

$y = -4 \sin \frac{5\pi}{6} = (-4)\left(\frac{1}{2}\right) = -2$.

Thus P: $\left(2\sqrt{3}, -2\right)$.

B. The procedure for graphing a polar equation is the same as that you used when you first graphed functions -- compute values and plot points. The graphs of some polar equations are easily identifiable when the equation is transformed into rectangular coordinates. For example, $r = 6 \sin \theta$ becomes $r^2 = 6r \sin \theta$ or $x^2 + y^2 = 6y$. This is $x^2 + y^2 - 6y = 0$ or $x^2 + (y - 3)^2 = 9$, the circle of radius 3 centered at $(0, 3)$.

For a polar curve $r = f(\theta)$, we can switch to rectangular coordinates $x = f(\theta) \cos \theta$, $y = f(\theta) \sin \theta$. Treating θ as a parameter,

$$\frac{dy}{dx} = \frac{\dfrac{dy}{d\theta}}{\dfrac{dx}{d\theta}}$$

This is the slope of the tangent line to $r = f(\theta)$ at (r, θ).

6) Sketch a graph of $r = \sin \theta - \cos \theta$.

Compute several points for values of θ

(r, θ)

A: $(-1, 0)$

B: $\left(\dfrac{1}{2} - \dfrac{\sqrt{3}}{2} , \dfrac{\pi}{6} \right)$

C: $(0, \dfrac{\pi}{4})$

D: $(1, \dfrac{\pi}{2})$

E: $(\sqrt{2}, \dfrac{3\pi}{4})$

F: $(1, \pi)$ (same as A)

G: $\left(\dfrac{-1}{2} + \dfrac{\sqrt{3}}{2} , \dfrac{7\pi}{6} \right)$ (same as B)

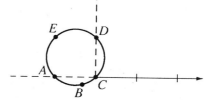

We can see the figure is a circle by switching to rectangular coordinates:

Multiply $r = \sin \theta - \cos \theta$ by r:

$r^2 = r \sin \theta - r \cos \theta$

$x^2 + y^2 = y - x$

$x^2 + x + y^2 - y = 0$

Completing the square gives

$$\left(x + \frac{1}{2}\right)^2 + \left(y - \frac{1}{2}\right)^2 = \frac{1}{2},$$

the circle with center $\left(-\frac{1}{2}, \frac{1}{2}\right)$ and radius $\frac{1}{\sqrt{2}}$.

7) Find the slope of the tangent line to

$r = e^\theta$ at $\theta = \frac{\pi}{2}$.

$x = r \cos \theta = e^\theta \cos \theta$, so

$\frac{dx}{d\theta} = e^\theta(- \sin \theta) + e^\theta(\cos \theta).$

$y = r \sin \theta = e^\theta \sin \theta$, so

$\frac{dy}{d\theta} = e^\theta \cos \theta + e^\theta \sin \theta.$

At $\theta = \frac{\pi}{2}$, $\frac{dx}{d\theta} = -e^{\pi/2}$ and $\frac{dy}{d\theta} = e^{\pi/2}$.

Therefore, $\frac{dy}{dx} = \dfrac{\dfrac{dy}{d\theta}}{\dfrac{dx}{d\theta}} = \dfrac{e^{\pi/2}}{-e^{\pi/2}} = -1.$

Areas and Lengths in Polar Coordinates

Concepts to Master

A. Area of a region described by polar equations
B. Length of a curve described by a polar equation

Summary and Focus Questions

A. The area of a region bounded by $\theta = a$, $\theta = b$, $r = f(\theta)$ where f is continuous and positive and

$$0 \le b - a \le 2\pi \text{ is } \int_a^b \frac{1}{2}[f(\theta)]^2 \, d\theta.$$

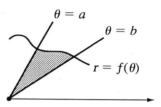

1) Find a definite integral for each region:
   a) the shaded area

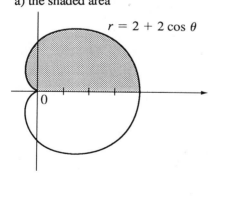

$$\text{Area} = \int_0^\pi \frac{1}{2}(2 + 2\cos\theta)^2 d\theta$$

$$= \int_0^\pi 2(1 + \cos\theta)^2 d\theta.$$

b) the region bounded by $\theta = \frac{\pi}{3}$, $\theta = \frac{\pi}{2}$, $r = e^\theta$.

$$\text{Area} = \int_{\pi/3}^{\pi/2} e^{2\theta}\, d\theta.$$

c) the shaded area

$r = 1 - \cos\theta$

$r = \cos\theta$

The area is between 2 curves so we must first determine where the curves intersect:

$1 - \cos\theta = \cos\theta$, $1 = 2\cos\theta$,

$\cos\theta = \dfrac{1}{2}$. Therefore $\theta = \dfrac{\pi}{3}, -\dfrac{\pi}{3}$.

For $-\dfrac{\pi}{3} \le \theta \le \dfrac{\pi}{3}$,

$\cos\theta \ge 1 - \cos\theta$ so the area is

$$= \int_{-\pi/3}^{\pi/3} \frac{1}{2}[\cos\theta]^2 d\theta - \int_{-\pi/3}^{\pi} \frac{1}{2}(1 - \cos\theta)^2 d\theta$$

$$= \int_{-\pi/3}^{\pi/3} \left(-\frac{1}{2} + \cos\theta\right) d\theta$$

B. A curve given in polar coordinates by $r = f(\theta)$ for $a \le \theta \le b$ has arc length

$$\int_a^b \sqrt{[f(\theta)]^2 + [f'(\theta)]^2}\, d\theta.$$

2) Set up a definite integral for the length
of each curve

a) the curve $r = e^{2\theta}$ for $0 \le \theta \le 1$.

$$\int_0^1 \sqrt{(e^{2\theta})^2 + (2e^{2\theta})^2} \, d\theta$$

$$= \sqrt{5} \int_0^1 e^{2\theta} \, d\theta.$$

b) the curve sketched below

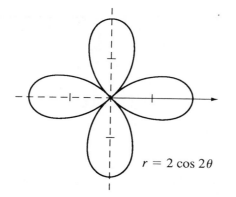

$r = 2 \cos 2\theta$

Since the curve is quite symmetric, the arc
length is 8 times the length of half of one leaf:

$$8 \int_0^{\pi/4} \sqrt{[2 \cos 2\theta]^2 + [-4 \sin 2\theta]^2} \, d\theta$$

$$= 8 \int_0^{\pi/4} \sqrt{4 \cos^2 2\theta + 16 \sin^2 2\theta} \, d\theta$$

$$= 16 \int_0^{\pi/4} \sqrt{1 + 3 \sin^2 2\theta} \, d\theta.$$

Conic Sections

Concepts to Master

Definitions, equations, properties, and graphs of

A. parabolas
B. ellipses
C. hyperbolas
D. equations of shifted conics

Summary and Focus Questions

A. Given a fixed point F and a fixed line D a <u>parabola</u> is the set of all points P such that the distance from F to P equals the distance from D to P.

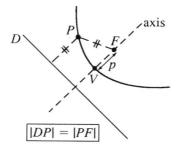

$|DP| = |PF|$

F is the <u>focus.</u>
D is the <u>directrix.</u>
The <u>axis</u> is the line perpendicular to D through F.
The <u>vertex</u> V is the point on the parabola midway between the focus and directrix.

The two basic forms for the equation of the parabola are given in the table on the following page. The vertex is situated at $(0, 0)$ and the distance from V to F is $|p|$. For each form there are two possible graphs, depending on whether p is positive or negative.

Equation   Axis   Focus   Directrix   Graphs

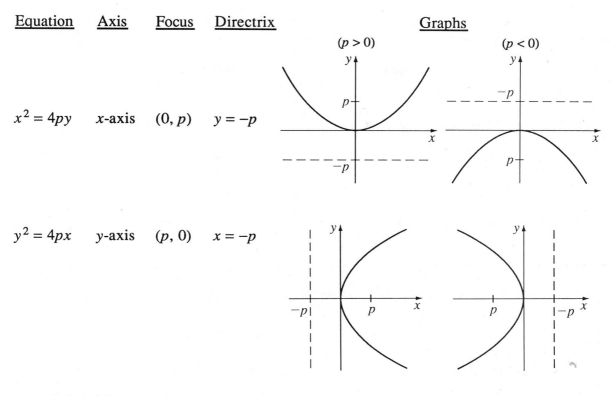

$x^2 = 4py$ x-axis $(0, p)$ $y = -p$

$y^2 = 4px$ y-axis $(p, 0)$ $x = -p$

1) Label the graph of the parabola.

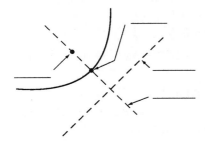

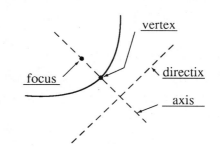

2) Graph $x^2 = -8y$, labeling the focus and directrix.

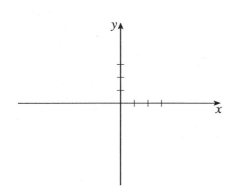

This parabola has the form $x^2 = 4py$ with $p = -2$.

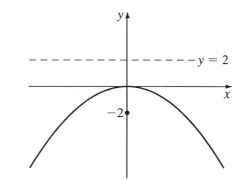

B. Given the fixed points F_1 and F_2, an <u>ellipse</u> is the set of all points P such that the distance from F_1 to P plus the distance from F_2 to P is constant.

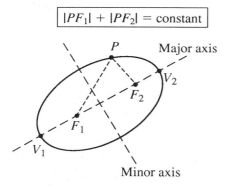

The points F_1 and F_2 are <u>foci</u> (plural of focus).
The line determined by F_1 and F_2 is the <u>major axis</u>.
The <u>minor axis</u> is the perpendicular bisector of the line segment $\overline{F_1 F_2}$.
The points V_1 and V_2 of intersection of the ellipse and the major axis are <u>vertices</u>.

If we place the foci F_1 and F_2 each at a distance c from $(0, 0)$ along an axis, let $2a$ be the constant $\overline{PF_1} + \overline{PF_2}$, and let $b = \sqrt{a^2 - c^2}$ (hence $a \geq b$), then the two basic forms for the equation of an ellipse are as follows.

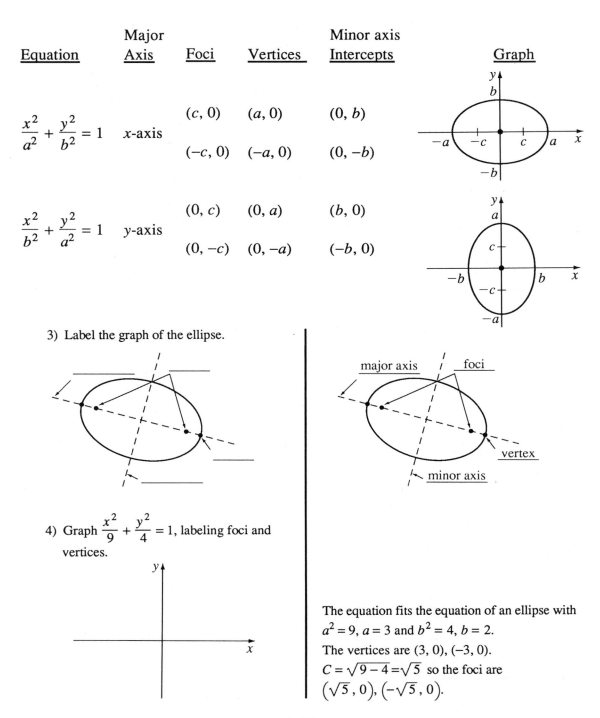

Equation	Major Axis	Foci	Vertices	Minor axis Intercepts	Graph
$\dfrac{x^2}{a^2} + \dfrac{y^2}{b^2} = 1$	x-axis	$(c, 0)$ $(-c, 0)$	$(a, 0)$ $(-a, 0)$	$(0, b)$ $(0, -b)$	
$\dfrac{x^2}{b^2} + \dfrac{y^2}{a^2} = 1$	y-axis	$(0, c)$ $(0, -c)$	$(0, a)$ $(0, -a)$	$(b, 0)$ $(-b, 0)$	

3) Label the graph of the ellipse.

major axis   foci

vertex

minor axis

4) Graph $\dfrac{x^2}{9} + \dfrac{y^2}{4} = 1$, labeling foci and vertices.

The equation fits the equation of an ellipse with $a^2 = 9$, $a = 3$ and $b^2 = 4$, $b = 2$.

The vertices are $(3, 0)$, $(-3, 0)$.

$C = \sqrt{9 - 4} = \sqrt{5}$ so the foci are $\left(\sqrt{5}, 0\right)$, $\left(-\sqrt{5}, 0\right)$.

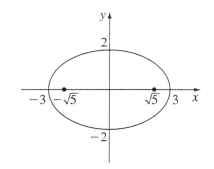

5) Find the equation of this ellipse:

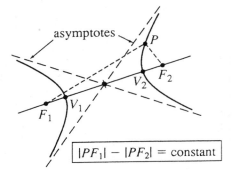

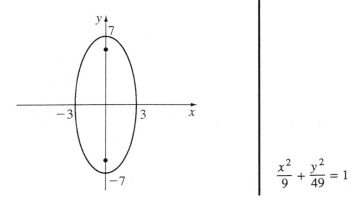

$$\frac{x^2}{9} + \frac{y^2}{49} = 1$$

C. Given two fixed points F_1 and F_2, a hyperbola is the set of all points P such that the distance from F_1 to P minus the distance from F_2 to P is constant.

F_1 and F_2 are <u>foci</u>.
The points V_1 and V_2 of intersection of the hyperbola and the line through F_1 and F_2 are <u>vertices</u>.
The <u>asymptotes</u> are two lines which the branches of the hyperbola approach.

$$|PF_1| - |PF_2| = \text{constant}$$

If we place the foci F_1 and F_2 each at a distance c from $(0, 0)$ along an axis, let $2a$ be the constant $\overline{PF_1} - \overline{PF_2}$, and let $b = \sqrt{c^2 - a^2}$, then the two basic forms for the equation of a hyperbola are as follows:

Equation	Foci	Vertices	Asymptotes	Graph

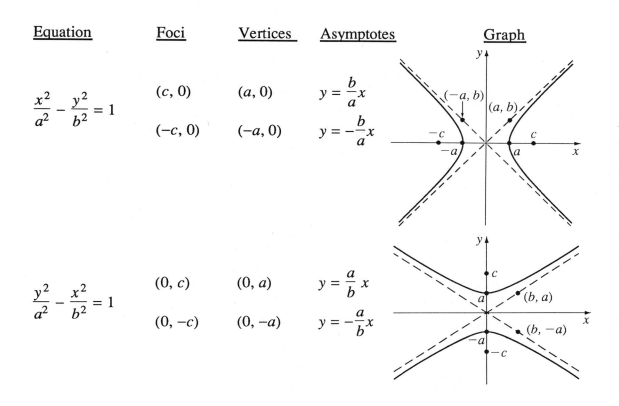

$\dfrac{x^2}{a^2} - \dfrac{y^2}{b^2} = 1$ $(c, 0)$ $(a, 0)$ $y = \dfrac{b}{a}x$

   $(-c, 0)$ $(-a, 0)$ $y = -\dfrac{b}{a}x$

$\dfrac{y^2}{a^2} - \dfrac{x^2}{b^2} = 1$ $(0, c)$ $(0, a)$ $y = \dfrac{a}{b}x$

   $(0, -c)$ $(0, -a)$ $y = -\dfrac{a}{b}x$

6) Label the graph of the hyperbola.

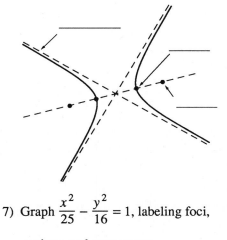

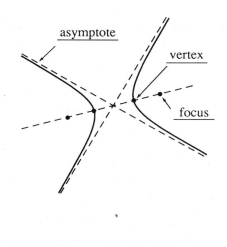

asymptote

vertex

focus

7) Graph $\dfrac{x^2}{25} - \dfrac{y^2}{16} = 1$, labeling foci,

vertices, and asymptotes.

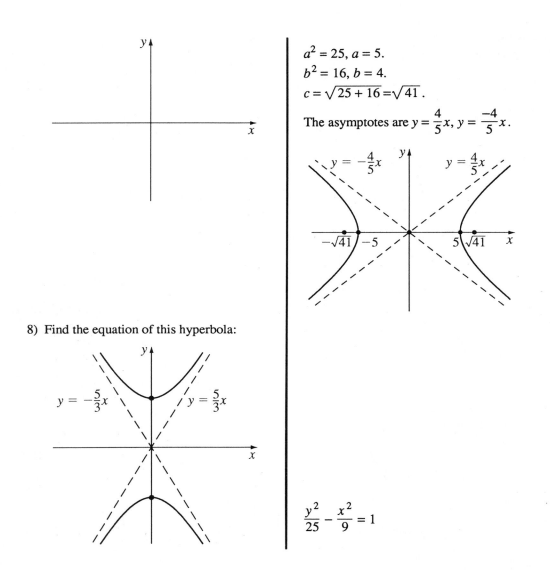

$a^2 = 25$, $a = 5$.

$b^2 = 16$, $b = 4$.

$c = \sqrt{25 + 16} = \sqrt{41}$.

The asymptotes are $y = \dfrac{4}{5}x$, $y = \dfrac{-4}{5}x$.

8) Find the equation of this hyperbola:

$$\frac{y^2}{25} - \frac{x^2}{9} = 1$$

D. Replacing x by $x - h$ and y by $y - k$ in any of the standard forms of a conic will shift the conic h units to the left and k units upward.
Given a second-degree equation of the form $Ax^2 + Bx + Cy^2 + Dy + E = 0$, $A \neq 0$ or $C \neq 0$, completing the square for x or y will transform the equation into one of the shifted basic forms.

9) Determine the type of conic and find the vertex for each:

a) $x^2 + 4x + 8y = 12$

There is no y^2 term so the equation represents a <u>parabola</u>.

$x^2 + 4x + 8y = 12$

$x^2 + 4x + 4 + 8y - 16 = 0$

$(x + 2)^2 + 8(y - 2) = 0$

$(x + 2)^2 = -8(y - 2)$.

The vertex is $(-2, 2)$.

b) $x^2 + 6x + 4y^2 - 4y = 12$

x^2 and y^2 both have the same sign (positive) so this is an <u>ellipse</u>.

$x^2 + 6x + 4y^2 - 4y = 12$

$x^2 + 6x + 9 + 4y^2 - 4y + 1 = 12 + 13$

$(x + 3)^2 + 4\left(y - \dfrac{1}{2}\right)^2 = 25$

$$\frac{(x + 3)^2}{25} + \frac{\left(y - \dfrac{1}{2}\right)^2}{25/4} = 1$$

For this ellipse $a = 5$.

The vertices are $\left(-3 \pm 5, \dfrac{1}{2}\right) = \left(-8, \dfrac{1}{2}\right), \left(2, \dfrac{1}{2}\right)$.

Conic Sections in Polar Coordinates

Concepts to Master

Focus-directrix definitions of conics; Eccentricity; Equations of conics in polar form

Summary and Focus Questions

Let F be a fixed point (focus), l be a fixed line (directrix), and e be a positive constant* (<u>eccentricity</u>). The set of all points P such that $\dfrac{|PF|}{|Pl|} = e$ is a conic.

<u>Eccentricity</u>	<u>Type</u>	<u>Graph</u>	
$e = 1$	parabola		$\dfrac{|PF|}{|Pl|} = 1$
$e < 1$	ellipse		$\dfrac{|PF|}{|Pl|} = e < 1$

*This use of e for eccentricity is not to be confused with the use of e as the symbol for the base of natural logarithms.

$e > 1$ $\qquad$ hyperbola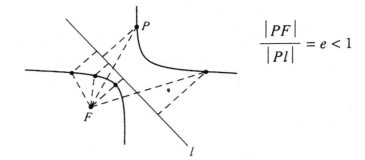

$$\frac{|PF|}{|Pl|} = e < 1$$

If a conic with eccentricity $e > 0$ is drawn in a rectangular coordinate system with focus F at $(0, 0)$ and directrix l one of the lines $x = d$, $x = -d$, $y = d$, $y = -d$, where $d > 0$, then by superimposing a polar coordinate system the polar equation of the conic has one of these forms:

<u>Conic</u> $\qquad\qquad\qquad$ Polar Form of Equation

$$r = \frac{ed}{1 + e \cos \theta} \qquad r = \frac{ed}{1 - e \cos \theta} \qquad r = \frac{ed}{1 + e \sin \theta} \qquad r = \frac{ed}{1 - e \sin \theta}$$

$\qquad$ ($x = d$ directrix) $\qquad$ ($x = -d$ directrix) $\qquad$ ($y = d$ directrix) $\qquad$ ($y = -d$ directrix)

parabola
$(e = 1)$

ellipse
$(0 < e < 1)$

hyperbola
$(e > 1)$

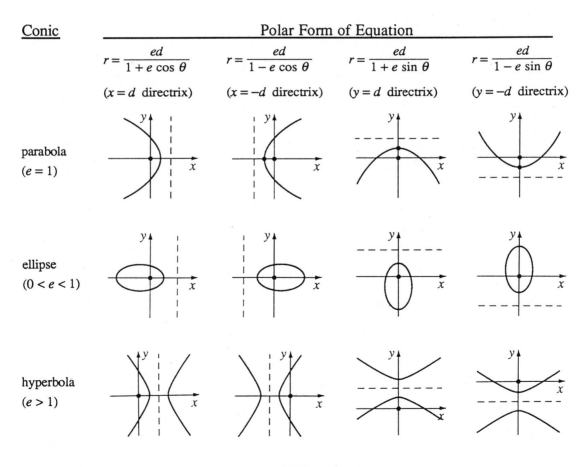

1) The sketch below shows one point P on a conic and its distance from the focus and directrix. What type of conic is it?

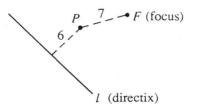

$$e = \frac{|PF|}{|Pl|} = \frac{7}{6} > 1, \text{ so the conic is a } \underline{\text{hyperbola}}.$$

2. What is the polar equation of the conic with

a) eccentricity 2, directrix $y = 3$.

$$e = 2 \text{ and } d = 3, \text{ so } r = \frac{6}{1 + 2 \sin \theta}.$$

b) eccentricity $\frac{1}{2}$, directrix $y = -3$.

$$e = \frac{1}{2}, \text{ and } d = 3, \text{ so } r = \frac{3/2}{1 - \frac{1}{2} \cos \theta},$$

$$r = \frac{3}{2 - \cos \theta}.$$

c) directrix $x = 4$ and is a parabola.

$$e = 1 \text{ and } d = 4, \text{ so } r = \frac{4}{1 + \cos \theta}.$$

3) Find the eccentricity and directrix and identify the conic given by

$$r = \frac{3}{4 - 5 \cos \theta} \quad .$$

Divide numerator and denominator by 4.

$$r = \frac{3/4}{1 - \frac{5}{4} \cos \theta}, \quad e = \frac{5}{4}. \text{ Since } ed = \frac{3}{4}$$

$$d = \frac{3}{4} \cdot \frac{1}{e} = \frac{3}{4} \cdot \frac{4}{5} = \frac{3}{5}.$$

The trigonometric term is $-\cos \theta$ so the directrix is

$$x = -\frac{3}{5}.$$

Since $e > 1$, the conic is a <u>hyperbola</u>.

4) What polar form does the equation of the ellipse below have?

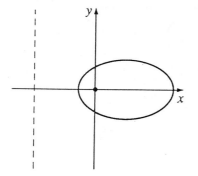

The directrix has the form $x = -d$ so the equation is $r = \dfrac{ed}{1 - e \cos \theta}$.

"I'M BEGINNING TO UNDERSTAND ETERNITY,
BUT INFINITY IS STILL BEYOND ME."

Sequences

Concepts to Master

Sequences; Limit of a sequence; Convergent; Divergent; Monotone sequences (increasing, decreasing); Bounded sequences

Summary and Focus Questions

A <u>sequence</u> is a set of numbers arranged as

$$\{a_n\} = a_1, a_2, a_3, \ldots, a_n, a_{n+1}, \ldots .$$

a_n is the <u>nth term</u> of the sequence.

The sequence $\{a_n\}$ <u>converges to L</u> means intuitively as n grows larger the a_n values get closer and closer to L. Formally, $\lim\limits_{n \to \infty} a_n = L$ means for all $\varepsilon > 0$ there exists a positive integer N such that $|a_n - L| < \varepsilon$ for all $n > N$. If the limit does not exist, $\{a_n\}$ <u>diverges.</u>

$\lim\limits_{n \to \infty} a_n = \infty$ means the a_n terms grow without bound.

One way to evaluate $\lim\limits_{n \to \infty} a_n$ is to find a real function $f(x)$ such that $f(n) = a_n$ for all n. If $\lim\limits_{x \to +\infty} f(x) = L$, then $\lim\limits_{n \to \infty} a_n = L$. The converse is false.

All limit laws for limits of functions at infinity are valid for convergent sequences. A version of the Squeeze Theorem also holds:

If $a_n \leq b_n \leq c_n$ for all $n \geq N$ and $\lim\limits_{n \to \infty} a_n = \lim\limits_{n \to \infty} c_n = L$, then $\lim\limits_{n \to \infty} b_n = L$.

$\{a_n\}$ is <u>increasing</u> means $a_{n+1} \geq a_n$ for all n.

$\{a_n\}$ is <u>decreasing</u> means $a_{n+1} \leq a_n$ for all n.

$\{a_n\}$ is <u>monotonic</u> if it is either increasing or decreasing.

One way to show $\{a_n\}$ is increasing is to show that $\dfrac{da_n}{dn} \geq 0$ (treating n as a real number).

$\{a_n\}$ is <u>bounded above</u> means $a_n \leq M$ for some M and all n.

$\{a_n\}$ is <u>bounded below</u> means $a_n \geq m$ for some m and all n.

$\{a_n\}$ is <u>bounded</u> if it is both bounded above and bounded below.

If $\{a_n\}$ converges, then $\{a_n\}$ is bounded. The converse is false.

If $\{a_n\}$ is bounded and monotonic, then $\{a_n\}$ converges. The converse is false.

1) Find the fourth term of the sequence
$$a_n = \frac{(-1)^n}{n^2}.$$

$$a_4 = \frac{(-1)^4}{4^2} = \frac{1}{16}.$$

2) Determine whether each converges

   a) $a_n = \dfrac{n}{n^2 + 1}.$

$$\lim_{n \to \infty} \frac{n}{n^2 + 1} = (\text{divide by } n^2)$$

$$\lim_{n \to \infty} \frac{\dfrac{1}{n}}{1 + \dfrac{1}{n^2}} = \frac{0}{1 + 0} = 0.$$

Thus a_n converges to 0.

   b) $b_n = \dfrac{n^2 + 1}{2n}$

b_n increases without bound so $\{b_n\}$ diverges.

We may write $\displaystyle\lim_{n \to \infty} \frac{n^2 + 1}{2n} = \infty.$

c) $c_n = \dfrac{(-1)^n n}{n+1}$.

The sequence $\dfrac{n}{n+1}$ converges to 1 but

$c_n = \dfrac{(-1)^n n}{n+1}$ is alternating so $\{c_n\}$ diverges.

3) Find $\lim\limits_{n \to \infty} \dfrac{\cos n}{n}$.

0. Since $-1 \le \cos n \le 1$, as $n \to \infty$, $\dfrac{\cos n}{n} \to 0$.

4) Find $\lim\limits_{n \to \infty} \dfrac{\ln n}{n}$.

Let $f(x) = \dfrac{\ln x}{x}$.

$\lim\limits_{x \to \infty} \dfrac{\ln x}{x} \left(= \dfrac{0}{0} \text{ form} \right)$

$= \lim\limits_{x \to \infty} \dfrac{1/x}{1}$ (L'Hospital's Rule) $= 0$.

Thus $\lim\limits_{x \to \infty} \dfrac{\ln n}{n} = 0$.

5) Is $c_n = \dfrac{1}{3n}$ bounded above? bounded below?

$\left\{ \dfrac{1}{3n} \right\}$ is bounded above by $\dfrac{1}{3}$ and bounded

below by 0.

6) Is $s_n = \dfrac{n}{n+1}$ an increasing sequence?

Yes, because $\dfrac{ds_n}{dn} = \dfrac{(n+1) - n}{(n+1)^2} = \dfrac{1}{(n+1)^2} > 0$.

$3(\wedge) \angle 5$

$3 \wedge \vee$

7) True or False:

a) If $\{a_n\}$ is not bounded below then $\{a_n\}$ diverges.

True.

b) If $\{a_n\}$ is decreasing and $a_n \geq 0$ for all n then $\lim\limits_{n \to \infty} a_n$ exists.

True.

c) If $\{a_n\}$ and $\{b_n\}$ converge then $\{a_n + b_n\}$ converges.

True.

d) If $\{a_n\}$ and $\{b_n\}$ diverge then $\{a_n + b_n\}$ diverges.

False. For example, $a_n = n^2$ and $b_n = -n^2$ diverge.

8) If $\lim\limits_{n \to \infty} s_n = 4$ and $\lim\limits_{n \to \infty} t_n = 2$, then

a) $\lim\limits_{n \to \infty} (8s_n - 2t_n) = $ _____ .

$8(4) - 2(2) = 28.$

b) $\lim\limits_{n \to \infty} \dfrac{3s_n}{t_n} = $ _____ .

$\dfrac{3(4)}{2} = 6.$

Series

Concepts to Master

A. Infinite series; Partial sums; Convergent and divergent series; Convergent series laws.
B. Geometric series; Value of a converging geometric series; Harmonic series

Summary and Focus Questions

A. $\sum_{k=1}^{\infty} a_n = a_1 + a_2 + a_3 + \ldots + a_n + \ldots$ is called an (infinite) series.

$s_n = \sum_{k=1}^{n} a_k = a_1 + a_2 + a_3 + \ldots + a_n + \ldots$ is the nth partial sum of the series.

$\sum_{n=1}^{\infty} a_n$ converges (to s) means $\lim_{n \to \infty} s_n$ exists (and is s). Thus a series converges if the limit of its sequence of partial sums exists.

If that limit does not exist, then $\sum_{n=1}^{\infty} a_n$ diverges.

If $\sum_{n=1}^{\infty} a_n$ converges, then $\lim_{n \to \infty} a_n = 0$, but not conversely.

Rewritten in contrapositive form this is the Test for Divergence:

If $\lim_{n \to \infty} a_n \neq 0$, $\sum_{k=1}^{\infty} a_k$ diverges.

If $\lim\limits_{n \to \infty} a_n = 0$, $\sum\limits_{n=1}^{\infty} a_n$ may converge or may diverge.

If $\sum\limits_{n=1}^{\infty} a_n$ converges (to L) and $\sum\limits_{n=1}^{\infty} b_n$ converges (to M) then:

$\sum\limits_{n=1}^{\infty} (a_n \pm b_n)$ converges (to $L \pm M$).

$\sum\limits_{n=1}^{\infty} ca_n$ converges (to cL) for c any constant.

1) A series converges if the _____ of the
   sequence of _____ exists.

limit, partial sums

2) Find the first four partial sums of $\sum\limits_{n=1}^{\infty} \dfrac{1}{k^2}$.

$$s_1 = \frac{1}{1^2} = 1.$$

$$s_2 = \frac{1}{1^2} + \frac{1}{2^2} = 1 + \frac{1}{4} = \frac{5}{4}.$$

$$s_3 = \frac{1}{1^2} + \frac{1}{2^2} + \frac{1}{3^2} = \frac{5}{4} + \frac{1}{9} = \frac{49}{36}.$$

$$s_4 = \frac{1}{1^2} + \frac{1}{2^2} + \frac{1}{3^2} + \frac{1}{4^2} = \frac{49}{36} + \frac{1}{16} = \frac{205}{144}.$$

3) Suppose $\sum\limits_{n=1}^{\infty} a_n = 3$ and $\sum\limits_{n=1}^{\infty} b_n = 4$.

   Evaluate each of the following:

a) $\displaystyle\sum_{n=1}^{\infty} (a_n + 2b_n)$

$3 + 2(4) = 11.$

b) $\displaystyle\sum_{n=1}^{\infty} \frac{a_n}{5}$

$\dfrac{3}{5}$

c) $\displaystyle\sum_{n=1}^{\infty} \frac{1}{a_n}$

This <u>diverges</u> because if $\displaystyle\sum_{n=1}^{\infty} a_n = 3$, then

$\displaystyle\lim_{n \to \infty} a_n = 0$. Thus $\displaystyle\lim_{n \to \infty} \frac{1}{a_n} \neq 0$, so $\displaystyle\sum_{n=1}^{\infty} \frac{1}{a_n}$

diverges.

4) Sometimes, Always, or Never:

a) $\displaystyle\sum_{n=1}^{\infty} a_n$ converges, but $\displaystyle\lim_{n \to \infty} a_n$

does not exist.

<u>Never.</u> If $\displaystyle\sum_{n=1}^{\infty} a_n$ converges, then

$\displaystyle\lim_{n \to \infty} a_n$ exists and is 0.

b) $\displaystyle\sum_{n=1}^{\infty} a_n$ converges, $\displaystyle\sum_{n=1}^{\infty} (a_n + b_n)$

converges, but $\displaystyle\sum_{n=1}^{\infty} b_n$ diverges.

<u>Never.</u> $\displaystyle\sum_{n=1}^{\infty} b_n = \sum_{n=1}^{\infty} (a_n + b_n) - \sum_{n=1}^{\infty} a_n$

is the difference of two convergent series and must converge.

5) Does $\displaystyle\sum_{n=1}^{\infty} \cos\left(\frac{1}{n}\right)$ converge?

No, the nth term does not approach 0.

B. A <u>geometric series</u> has the form

$$\sum_{n=1}^{\infty} ar^{n-1} = a + ar + ar^2 + ar^3 + \ldots$$

and converges (for $a \neq 0$) to $\dfrac{a}{1-r}$ if and only if $-1 < r < 1$.

The <u>harmonic series</u> $\displaystyle\sum_{n=1}^{\infty} \frac{1}{n} = 1 + \frac{1}{2} + \frac{1}{3} + \ldots + \frac{1}{n} + \ldots$ diverges.

6) Sometimes, Always, or Never:

If $\displaystyle\lim_{n \to \infty} a_n = 0$, $\displaystyle\sum_{n=1}^{\infty} a_n$ converges.

<u>Sometimes.</u> True for $\displaystyle\sum_{n=1}^{\infty} \frac{1}{2^n}$ and false for

$$\sum_{n=1}^{\infty} \frac{1}{n}.$$

7) a) $\displaystyle\sum_{n=1}^{\infty} \frac{2^n}{100}$ is a geometric series in

which $a =$ _____ and
$r =$ _____.

$$\sum_{n=1}^{\infty} \frac{2^n}{100} = \frac{2}{100} + \frac{4}{100} + \frac{8}{100} + \dots$$

so $a = \dfrac{2}{100}$ and $r = 2$.

b) Does it converge?

No, since $r \geq 1$.

8) $\displaystyle\sum_{n=1}^{\infty} \left(\frac{-2}{9}\right)^n =$ _____.

A geometric series with $a = r = -\dfrac{2}{9}$

which converges to $\dfrac{a}{1-r} = -\dfrac{2}{11}$.

9) Does $\displaystyle\sum_{n=1}^{\infty} \frac{6}{n}$ converge?

It diverges, since it is a multiple of the harmonic series.

10) $\displaystyle\sum_{n=1}^{\infty} \frac{2^n}{3^{n+1}} =$

$$\sum_{n=1}^{\infty} \frac{2^n}{3^{n+1}} = \sum_{n=1}^{\infty} \frac{2}{9}\left(\frac{2}{3}\right)^{n-1}, \text{ which is a}$$

geometric series with $a = \dfrac{2}{9}$, $r = \dfrac{2}{3}$ that

converges to $\dfrac{2/9}{1 - 2/3} = \dfrac{2}{3}$.

11) Achilles gives the tortoise a 100 m head start. If Achilles runs at 5 m/sec and the tortoise at $\frac{1}{2}$ m/sec how far has the tortoise traveled by the time Achilles catches him ?

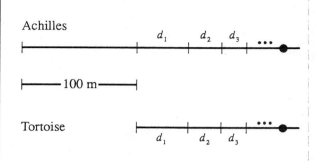

Achilles

Tortoise

Let d_1 = distance tortoise traveled while Achilles was running to the tortoise's starting point.

$$d_1 = \left(\frac{100\text{ m}}{5\text{ m/sec}}\right)\left(\frac{1}{2}\text{ m/sec}\right) = 10\text{ m}.$$

Let d_2 = distance tortoise traveled while Achilles was running the distance d_1.

Since $d_1 = 10$ and Achilles runs at 5 m/sec

$$d_2 = \left(\frac{10\text{ m}}{5\text{ m/sec}}\right)\frac{1}{2}\text{ m/sec} = 1\text{ m}.$$

For each n, $d_n = \left(\frac{d_{n-1}}{5}\right)\frac{1}{2} = \frac{d_{n-1}}{10}$.

The total distance traveled by the tortoise is

$$\sum_{n=1}^{\infty} d_n = 10 + 1 + \frac{1}{10} + \dots$$

which is a geometric series with $a = 10$, $r = \frac{1}{10}$.

This converges to $\dfrac{10}{1 - \dfrac{1}{10}} = \dfrac{100}{9}$ m.

The Integral Test

Concepts to Master

Integral Test for convergence; *p*-series

Summary and Focus Questions

Integral Test: Suppose that $f(x)$ is positive, continuous, and decreasing for $x \geq 1$ such that $a_n = f(k)$ for all n. Then $\sum_{n=1}^{\infty} a_n$ converges if and only if $\int_{1}^{\infty} f(x)\, dx$ converges. The Integral Test works when a_n has the form of a function whose antiderivative is easily found. It is one of the few general tests that gives necessary and sufficient conditions for convergence.

A *p*-series has the form $\sum_{n=1}^{\infty} \frac{1}{n^p}$, ($p$ a constant), and converges (by the Integral Test) if and only if $p > 1$.

1) True or False:

The Integral Test will determine to what value a series converges.

False. The test only indicates whether a series converges.

2) Test $\sum_{n=1}^{\infty} \frac{n}{e^n}$ for convergence.

$\frac{n}{e^n}$ suggests the function $f(x) = x\, e^{-x}$ which on

– 287 –

$[1, \infty)$ is continuous and decreasing.

$$\int_1^\infty x\,e^{-x} = \lim_{t \to \infty} \int_1^t x\,e^{-x}\,dx.$$

Using integration by parts,

$$\left(\begin{array}{ll} u = x & dv = e^{-x}\,dx \\ du = dx & dv = -e^{-x} \end{array} \right)$$

$$\int_1^t x\,e^{-x}\,dx = -x\,e^{-x} - e^{-x}\,\bigg|_1^t$$

$$= \frac{2}{e} - \frac{t+1}{e^t}.$$

$$\lim_{t \to \infty} \frac{2}{e} - \frac{t+1}{e^t} = \frac{2}{e} - 0 = \frac{2}{e}.$$

$$\left(\text{By L'Hospital's Rule, } \frac{t+1}{e^t} \to 0. \right)$$

Thus $\int_1^\infty x\,e^{-x}\,dx = \frac{2}{e}$, so $\sum_{n=1}^\infty \frac{n}{e^n}$ converges.

3) Which of these converge?

a) $\sum_{n=1}^\infty \frac{1}{n^2}$.

<u>Converges</u> (p-series with $p = 2$).

b) $\sum_{n=1}^\infty \frac{1}{\sqrt{n}}$.

<u>Diverges</u> (p-series with $p = \frac{1}{2}$).

-- 288 --

c) $\displaystyle\sum_{n=1}^{\infty} \frac{3}{2n^3}$.

Converges. $\displaystyle\sum_{n=1}^{\infty} \frac{3}{2n^3} = \frac{3}{2} \sum_{n=1}^{\infty} \frac{1}{n^3}$ is a

multiple of a p-series with $p = 3$.

d) $\displaystyle\sum_{n=1}^{\infty} \left(\frac{1}{n^3} + \frac{1}{8^n} \right)$.

Converges. $\displaystyle\sum_{n=1}^{\infty} \left(\frac{1}{n^3} + \frac{1}{8^n} \right) = \sum_{n=1}^{\infty} \frac{1}{n^3} + \sum_{n=1}^{\infty} \frac{1}{8^n}$,

the sum of a convergent p-series ($p = 3$) and a

convergent geometric series ($r = \frac{1}{8}$).

The Comparison Tests

Concepts to Master

Comparison Test; Limit Comparison Test

Summary and Focus Questions

The following tests may be used to determine the convergence of a series whose *terms are all positive*. They do not indicate to what number the series converges, only whether it converges or diverges.

Comparison Test:

1. If there is a convergent series $\sum\limits_{n=1}^{\infty} b_n$ such that $0 \le a_n \le b_n$ for all $n \ge N$,

then $\sum\limits_{n=1}^{\infty} a_n$ converges.

2. If there is a divergent series $\sum\limits_{n=1}^{\infty} b_n$ such that $0 \le b_n \le a_n$ for all $n \ge N$,

then $\sum\limits_{n=1}^{\infty} a_n$ diverges.

Limit Comparison Test:

Let $\sum\limits_{n=1}^{\infty} a_n$ and $\sum\limits_{n=1}^{\infty} b_n$ be positive term series.

1. If $\lim\limits_{n \to \infty} \dfrac{a_n}{b_n} = c > 0$, then $\sum\limits_{n=1}^{\infty} a_n$ and $\sum\limits_{n=1}^{\infty} b_n$ either both converge or both diverge.

2. If $\lim\limits_{n \to \infty} \dfrac{a_n}{b_n} = 0$, and $\sum\limits_{n=1}^{\infty} b_n$ converges, then $\sum\limits_{n=1}^{\infty} a_n$ converges.

3. If $\lim\limits_{n \to \infty} \dfrac{a_n}{b_n} = 0$, and $\sum\limits_{n=1}^{\infty} b_n$ diverges, then $\sum\limits_{n=1}^{\infty} a_n$ diverges.

To successfully use either test for a series $\sum\limits_{n=1}^{\infty} a_n$, you must come up with another series $\sum\limits_{n=1}^{\infty} b_n$ (geometric, p-series, ...) that you know is convergent or divergent and compare a_n to b_n.

1) Determine whether or not each of the following converge. Find a series $\sum\limits_{n=1}^{\infty} b_n$ to use with the Comparison Test or Limit Comparison Test.

a) $\sum\limits_{n=1}^{\infty} \dfrac{1}{n^2 + 2n}$, $b_n = $ _____.

You must have a hunch ahead of time whether $\sum\limits_{n=1}^{\infty} \dfrac{1}{n^2 + 2n}$ converges. Because $\dfrac{1}{n^2 + 2n}$ is "like" $\dfrac{1}{n^2}$ for large n, and $\sum\limits_{n=1}^{\infty} \dfrac{1}{n^2}$ is a converging p-series, our hunch is the given series converges.

Use $b_n = \dfrac{1}{n^2}$.

Since $\dfrac{1}{n^2 + 2n} < \dfrac{1}{n^2}$, $\displaystyle\sum_{n=1}^{\infty} \dfrac{1}{n^2 + 2n}$

converges by the Comparison Test.

b) $\displaystyle\sum_{n=1}^{\infty} \dfrac{\sqrt[3]{n}}{n+4}$, $b_n =$ _____.

Your hunch should be $\dfrac{\sqrt[3]{n}}{n+4}$ is "like"

$\dfrac{\sqrt[3]{n}}{n} = \dfrac{1}{\sqrt[3]{n^2}}$. Use $b_n = \dfrac{1}{\sqrt[3]{n^2}}$.

Then $\displaystyle\lim_{n\to\infty} \dfrac{a_n}{b_n} = \lim_{n\to\infty} \dfrac{\dfrac{\sqrt[3]{n}}{n+4}}{\dfrac{1}{\sqrt[3]{n^2}}} = \lim_{n\to\infty} \dfrac{n}{n+4} = 1.$

Since $\displaystyle\sum_{n=1}^{\infty} \dfrac{1}{\sqrt[3]{n^2}}$ diverges (a p-series with $p = 2/3$),

$\displaystyle\sum_{n=1}^{\infty} \dfrac{\sqrt[3]{n}}{n+4}$ diverges by the Limit Comparison Test.

c) $\displaystyle\sum_{n=1}^{\infty} \dfrac{1}{n + 2^n}$, $b_n =$ _____.

Use $b_n = \dfrac{1}{2^n}$. Then for $n \geq 1$, $\dfrac{1}{n + 2^n} \leq \dfrac{1}{2^n}$.

Since $\displaystyle\sum_{n=1}^{\infty} \frac{1}{2^n}$ converges (a geometric series with

$r = \frac{1}{2}$), $\displaystyle\sum_{n=1}^{\infty} \frac{1}{n + 2^n}$ converges.

2) Suppose $\{a_n\}$ and $\{b_n\}$ are positive sequences.

Sometimes, Always, Never:

a) If $\displaystyle\lim_{n \to \infty} \frac{a_n}{b_n} = 0$ and $\displaystyle\sum_{k=1}^{\infty} a_k$ converges,

then $\displaystyle\sum_{k=1}^{\infty} b_k$ converges.

<u>Sometimes.</u> For $a_n = \dfrac{1}{n^3}$ and $b_n = \dfrac{1}{n^2}$ it is true.

For $a_n = \dfrac{1}{n^2}$ and $b_n = \dfrac{1}{n}$ it is false.

b) If $\displaystyle\lim_{n \to \infty} \frac{a_n}{b_n} = \infty$ and $\displaystyle\sum_{n=1}^{\infty} b_n$

diverges, then $\displaystyle\sum_{n=1}^{\infty} a_n$ diverges.

<u>Always.</u>

Alternating Series

Concepts to Master

A. Alternating Series Tests
B. Estimating the value of a convergent alternating series

Summary and Focus Questions

A. An <u>alternating series</u> has successive terms of opposite signs, that is, has either the

form $\sum_{n=1}^{\infty} (-1)^n a_n$ or $\sum_{n=1}^{\infty} (-1)^{n+1} a_n$, where $a_n > 0$.

The <u>Alternating Series Test</u>: If $\{a_n\}$ is a decreasing sequence with $\lim_{n \to \infty} a_n = 0$, then

$\sum_{k=1}^{\infty} (-1)^n a_n$ converges.

1) Determine whether each converge:

a) $\sum_{n=1}^{\infty} \dfrac{(-1)^n}{e^n}$

This is an alternating series with $a_n = \dfrac{1}{e^n}$.

$\dfrac{1}{e^{n+}} \leq \dfrac{1}{e^n}$ and $\lim_{n \to \infty} \dfrac{1}{e^n} = 0.$

Thus $\sum_{n=1}^{\infty} \dfrac{(-1)^n}{e^n}$ <u>converges.</u>

b) $\displaystyle\sum_{n=1}^{\infty} \frac{(-1)^n n}{n+1}$

This is an alternating series but the other conditions for the test do not hold.

However $\displaystyle\lim_{n\to\infty} \frac{(-1)^n n}{n+1} \neq 0$, so $\displaystyle\sum_{n=1}^{\infty} \frac{(-1)^n n}{n+1}$

diverges.

B. If $\displaystyle\sum_{n=1}^{\infty} (-1)^n a_n$ is an alternating series with $0 \le a_{n+1} \le a_n$ and $\displaystyle\lim_{n\to\infty} a_n = 0$ then

$$|s_n - s| \le a_{n+1}.$$

Since the difference between the limit of the series and the nth partial sum does not exceed a_{n+1}, s_n may be used to approximate s to within an accuracy of a_{n+1}.

2) The series $\displaystyle\sum_{n=1}^{\infty} \frac{(-1)^n}{\sqrt{n}+1}$ converges.

Estimate the error between the sum of the series and its fifteenth partial sum.

Since the series alternates and $\dfrac{1}{\sqrt{n}+1}$

decreases to 0, the error $|s_{15} - s|$ does not exceed

$$a_{16} = \frac{1}{\sqrt{16}+1} = .2 \ .$$

3) For what value of n is s_n, the nth partial

sum, within .001 of $s = \displaystyle\sum_{n=1}^{\infty} \frac{(-1)^{n+1}}{2n+5}$?

$$\left| s_n - s \right| \le a_{n+1} = \frac{1}{(2n+1)+5} \le .001$$

$2n + 6 \ge 1000,\ 2n \ge 994,\ n \ge 497.$

Let $n = 497$. Then s_{497} is within .001 of s.

4) Approximate $\displaystyle\sum_{n=1}^{\infty} \frac{(-1)^{n+1}}{n^3}$ to within .005.

$$\left| s_n - s \right| \le a_{n+1} = \frac{1}{(n+1)^3} \le .005.$$

$$(n+1)^3 \ge \frac{1}{.005} = 200.$$

For $n = 5$, $(n+1)^3 = 6^3 = 216 > 200$.

Therefore s_5 is within .005 of s.

$$s_5 = \frac{1}{1^3} - \frac{1}{2^3} + \frac{1}{3^3} - \frac{1}{4^3} + \frac{1}{5^3}.$$

$$= 1 - \frac{1}{8} + \frac{1}{27} - \frac{1}{64} + \frac{1}{125}$$

$$\approx .9044.$$

We do not know to what the series converges, but that value is within .005 of .9044.

Absolute Convergence and the Ratio and Root Tests

Concepts to Master

A. Absolute convergence; Conditional convergence
B. Ratio Test; Root Test

Summary and Focus Questions

A. For any series (with a_n not necessarily positive or alternating) the concept of convergence may be split into two subconcepts:

$$\sum_{k=1}^{\infty} a_k \text{ converges absolutely if } \sum_{k=1}^{\infty} |a_k| \text{ converges.}$$

$$\sum_{k=1}^{\infty} a_k \text{ converges conditionally if } \sum_{k=1}^{\infty} a_k \text{ converges and } \sum_{k=1}^{\infty} |a_k| \text{ diverges.}$$

Either one of absolute convergence or conditional convergence implies (ordinary) convergence. Conversely, convergence implies exactly one of absolute or conditional convergence. Thus *every* series must do precisely one of these three: diverge, converge absolutely, or converge conditionally.

1) Sometimes, Always, or Never:

a) If $\sum_{n=1}^{\infty} |a_n|$ diverges, then $\sum_{n=1}^{\infty} a_n$

diverges.

<u>Sometimes.</u> True for $a_n = n$ but false for

$$a_n = \frac{(-1)^n}{n}.$$

b) If $\displaystyle\sum_{n=1}^{\infty} a_n$ diverges, then $\displaystyle\sum_{n=1}^{\infty} |a_n|$

diverges.

<u>Always.</u>

c) If $a_n \geq 0$ for all n, then $\displaystyle\sum_{n=1}^{\infty} a_n$ is

not conditionally convergent.

<u>Always.</u>

2) Determine whether each converges
conditionally or absolutely, or diverges.

a) $\displaystyle\sum_{n=1}^{\infty} \frac{(-1)^n}{\sqrt{n}}$

The series is alternating with $\dfrac{1}{\sqrt{n}}$

decreasing to 0 so it converges. It remains
to check for absolute convergence:

$$\sum_{n=1}^{\infty} \left| \frac{(-1)^n}{\sqrt{n}} \right| = \sum_{n=1}^{\infty} \frac{1}{\sqrt{n}} \text{ diverges}$$

(p-series with $p = \dfrac{1}{2}$). Thus $\displaystyle\sum_{n=1}^{\infty} \frac{(-1)^n}{\sqrt{n}}$

converges conditionally.

b) $\displaystyle\sum_{n=1}^{\infty} \frac{\sin n + \cos n}{n^3}$

Check for absolute convergence first:

$$\sum_{n=1}^{\infty} \left| \frac{\sin n + \cos n}{n^3} \right| = \sum_{n=1}^{\infty} \frac{|\sin n + \cos n|}{n^3}.$$

Since $|\sin n + \cos n| \leq 2$ and $\displaystyle\sum_{n=1}^{\infty} \frac{2}{n^3}$

converges (it is a p-series with $p = 3$),

$\displaystyle\sum_{n=1}^{\infty} \frac{|\sin n + \cos n|}{n^3}$ converges by the

Comparison Test. Thus $\displaystyle\sum_{n=1}^{\infty} \frac{\sin n + \cos n}{n^3}$

converges absolutely.

3) Does $\displaystyle\sum_{n=1}^{\infty} \frac{\sin e^n}{e^n}$ converge?

Yes. The series is not alternating but does contain both positive and negative terms. Check for absolute convergence:

$$\sum_{n=1}^{\infty} \left| \frac{\sin e^n}{e^n} \right| = \sum_{n=1}^{\infty} \frac{|\sin e^n|}{e^n}.$$

Since $|\sin e^n| \leq 1$,

$$\frac{|\sin e^n|}{e^n} \leq \frac{1}{e^n}.$$

$\displaystyle\sum_{n=1}^{\infty} \frac{1}{e^n}$ converges (geometric series with $r = \frac{1}{e}$) so

by the Comparison Test $\displaystyle\sum_{n=1}^{\infty} \left| \frac{\sin e^n}{e^n} \right|$

converges. Therefore $\displaystyle\sum_{n=1}^{\infty} \frac{\sin e^n}{e^n}$ converges

absolutely and hence converges.

Section 10.6: Absolute Convergence and the Ratio and Root Tests

B. Let a_n be a sequence of nonzero terms. The following may be used to determine whether $\displaystyle\sum_{n=1}^{\infty} a_n$ converges.

Ratio Test:

   1. If $\displaystyle\lim_{n\to\infty} \left| \frac{a_{n+1}}{a_n} \right| = L < 1$, then $\displaystyle\sum_{n=1}^{\infty} a_n$ converges absolutely (and

   therefore converges).

   2. If $\displaystyle\lim_{n\to\infty} \left| \frac{a_{n+1}}{a_n} \right| = L > 1$ or $\displaystyle\lim_{n\to\infty} \left| \frac{a_{n+1}}{a_n} \right| = \infty$, then $\displaystyle\sum_{n=1}^{\infty} a_n$ diverges.

If $\displaystyle\lim_{n\to\infty} \left| \frac{a_{n+1}}{a_n} \right| = 1$, $\displaystyle\sum_{n=1}^{\infty} a_n$ may converge or diverge.

Root Test:

   1. If $\displaystyle\lim_{n\to\infty} \sqrt[n]{|a_n|} = L < 1$, then $\displaystyle\sum_{n=1}^{\infty} a_n$ converges absolutely (and therefore

   converges).

   2. If $\displaystyle\lim_{n\to\infty} \sqrt[n]{|a_n|} = L > 1$, or $\displaystyle\lim_{n\to\infty} \sqrt[n]{|a_n|} = \infty$, then $\displaystyle\sum_{n=1}^{\infty} a_n$ diverges.

If $\displaystyle\lim_{n\to\infty} \sqrt[n]{|a_n|} = 1$, $\displaystyle\sum_{n=1}^{\infty} a_n$ may converge or diverge.

The Ratio and Root Tests involve no other series or functions and so are relatively easy to use, but fail for $L = 1$. The Ratio Test often works when a_n contains exponentials and/or factorials and will not work when a_n is a rational function of n. The Root Test works best when a_n contains an expression to the nth power.

4) Determine whether each converges by the Ratio Test:

a) $\displaystyle\sum_{n=1}^{\infty} \frac{n}{4^n}$

$$\lim_{n \to \infty} \left| \frac{a_{n+1}}{a_n} \right| = \lim_{n \to \infty} \frac{\dfrac{n+1}{4^{n+1}}}{\dfrac{n}{4^n}}$$

$$= \lim_{n \to \infty} \frac{n+1}{4n} = \frac{1}{4}. \text{ Thus } \sum_{n=1}^{\infty} \frac{n}{4^n}$$

converges absolutely and therefore converges.

b) $\displaystyle\sum_{n=1}^{\infty} \frac{(-4)^n}{n!}$

$$\lim_{n \to \infty} \left| \frac{a_{n+1}}{a_n} \right| = \lim_{n \to \infty} \left| \frac{\dfrac{(-4)^{n+1}}{(n+1)!}}{\dfrac{(-4)^n}{n!}} \right|$$

$$= \lim_{n \to \infty} \frac{4}{n+1} = 0. \text{ Thus } \sum_{n=1}^{\infty} \frac{(-4)^n}{n!} \text{ converges}$$

absolutely and therefore converges.

c) $\displaystyle\sum_{n=1}^{\infty} \frac{(-1)^n}{\sqrt[3]{n}}$

$$\lim_{n \to \infty} \left| \frac{a_{n+1}}{a_n} \right| = \lim_{n \to \infty} \left| \frac{\dfrac{(-1)^{n+1}}{\sqrt[3]{n+1}}}{\dfrac{(-1)^n}{\sqrt[3]{n}}} \right|$$

$$= \lim_{n \to \infty} \sqrt[3]{\frac{n}{n+1}} = 1. \text{ The Ratio Test \underline{fails}. The}$$

series does converge by the Alternating Series Test.

5) Determine whether each converges by the Root Test:

a) $\displaystyle\sum_{n=1}^{\infty} \frac{1}{(n+1)^n}$

$$\lim_{n \to \infty} \sqrt[n]{|a_n|} = \lim_{n \to \infty} \sqrt[n]{1/(n+1)^n}$$

$$= \lim_{n \to \infty} \frac{1}{n+1} = 0.$$

Thus $\displaystyle\sum_{n=1}^{\infty} \frac{1}{(n+1)^n}$ converges absolutely and

therefore converges.

b) $\displaystyle\sum_{n=1}^{\infty} \frac{3^n}{n^3}$

$$\lim_{n \to \infty} \sqrt[n]{|a_n|} = \lim_{n \to \infty} \sqrt[n]{\frac{3^n}{n^3}}$$

$$= \lim_{n \to \infty} \frac{3}{n^{3/n}} = \frac{3}{1} = 3.$$

Therefore by the Root Test $\displaystyle\sum_{n=1}^{\infty} \frac{3^n}{n^3}$ diverges.

$\left(\displaystyle\lim_{n \to \infty} n^{3/n} = 1\right.$ using the methods of Section 3.9:

Let $y = \displaystyle\lim_{n \to \infty} n^{3/n}$.

Then $\ln y = \displaystyle\lim_{n \to \infty} \ln n^{3/n}$

$= \displaystyle\lim_{n \to \infty} \frac{3 \ln n}{n}$ (form $\frac{\infty}{\infty}$)

$= \displaystyle\lim_{n \to \infty} \frac{3/n}{1} = 0.$

Thus $y = e^0 = 1.)$

Power Series

Concepts to Master

Power series in $(x - c)$; Interval of convergence; Radius of convergence

Summary and Focus Questions

A <u>power series in $(x - c)$</u> is an expression of the form $\sum\limits_{n=0}^{\infty} a_n (x - c)^n$, where a_n and c are constants. For any particular x, the value of the power series is an infinite series. For some values of x it will converge while it may diverge for other values.

The domain of a power series in $x - c$ is called its <u>interval of convergence</u> and contains those values of x for which the series converges. It consists of all real numbers from $c - R$ to $c + R$ for some R with $0 \leq R \leq \infty$. The number c is called the <u>center</u> and R is called the <u>radius</u> of convergence. When $R = 0$, the interval is the single point $\{c\}$. When $R = \infty$, the interval is $(-\infty, \infty)$.

The radius R is determined by using the Ratio Test on $\sum\limits_{n=0}^{\infty} a_n (x - c)^n$. When $0 < R < \infty$, the Ratio Test gives no information about the two endpoints $c - R$ and $c + R$. These endpoints may or may not be in the interval of convergence and must be tested separately using some other test. You should remember that the power series diverges for all x not in the interval of convergence.

1) True or False:

A power series is an infinite series.

False. A power series is an expression for a function which becomes an infinite series when a value is substituted for its variable.

2) For what value of x does $\displaystyle\sum_{n=0}^{\infty} a_n (x - c)^n$

always converge?

At $x = c$, $\displaystyle\sum_{n=0}^{\infty} a_n (x - c)^n = a_0$.

3) Compute the value of

a) $\displaystyle\sum_{n=0}^{\infty} n^2(x - 3)^n$ at $x = 4$.

When $x = 4$ the power series is

$$\sum_{n=0}^{\infty} n^2\, 1^n = \sum_{n=0}^{\infty} n^2.$$ This infinite series diverges

so there is no value for $x = 4$.

b) $\displaystyle\sum_{n=0}^{\infty} 2^n(x - 1)^n$ at $x = \frac{5}{6}$.

For $x = \frac{5}{6}$ $\displaystyle\sum_{n=0}^{\infty} 2^n\left(-\frac{1}{6}\right)^n = \sum_{n=0}^{\infty}\left(-\frac{1}{3}\right)^n$

$= 1 - \dfrac{1}{3} + \dfrac{1}{9} - \dfrac{1}{27} + \cdots$

(geometric series: $a = 1, r = -\dfrac{1}{3}$)

$= \dfrac{1}{1 - (-\frac{1}{3})} = \dfrac{3}{4}.$

$y' = \cos(\pi^3) \cdot 3\pi^2 - \sin(3x)^3 \cdot$

$\dfrac{7}{8} - \dfrac{8}{8}$

4) Find the center, radius, and interval of convergence for:

a) $\sum_{n=1}^{\infty} (2n)! \, (x-1)^n$

Use the Ratio Test

$$\lim_{n\to\infty} \left| \frac{(2(n+1))! \, (x-1)^{n+1}}{(2n)! \, (x-1)^n} \right|$$
$$= \lim_{n\to\infty} (2n+1)(2n+2) \, |x-1| = \infty,$$

except when $x=1$. Thus the center is 1, radius is 0, and interval of convergence is $\{1\}$.

b) $\sum_{n=1}^{\infty} \frac{(x-4)^n}{2^n n}$

$$\lim_{n\to\infty} \left| \frac{(x-4)^{n+1}}{2^{n+1}(n+1)} \cdot \frac{2^n n}{(x-4)^n} \right|$$

$$= \lim_{n\to\infty} \frac{n}{2(n+1)} \, |x-4| = \frac{1}{2}|x-4|.$$

Solving $\frac{1}{2}|x-4| < 1$: $|x-4| < 2$.

The center is 4 and radius is 2.
When $|x-4| = 2$, $x=2$ or $x=6$.

At $x=2$, the power series becomes $\sum_{n=1}^{\infty} \frac{(-1)^n}{n}$ which converges by the Alternating Series Test. At $x=6$ the power series is the divergent

harmonic series $\sum_{n=1}^{\infty} \frac{1}{n}$.

The interval of convergence is $[2, 6)$.

5) Suppose $\sum_{n=0}^{\infty} a_n(x-5)^n$ converges for $x=8$.

For what other values must it converge?

The interval of convergence is $5 - R$ to $5 + R$.

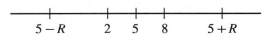

Because 8 is in this interval $8 \le 5 + R$. Thus $3 \le R$ which means the interval at least contains all numbers between $5 - 3$ and $5 + 3$. Thus the power series converges for at least all $x \in (2, 8]$.

Taylor and Maclaurin Series

Concepts to Master

A. Differentiation and integration of power series
B. Taylor series; Maclaurin series
C. Multiplication and division of power series

Summary and Focus Questions

A. If $f(x) = \sum\limits_{n=0}^{\infty} a_n (x - c)^n$ has interval of convergence with radius R then:

1) f is continuous on $(c - R, c + R)$.

2) For all $x \in (c - R, c + R)$, f may be differentiated term by term:

$$f'(x) = \sum_{n=1}^{\infty} n \, a_n (x - c)^{n-1}.$$

3) For all $x \in (c - R, c + R)$, f may be integrated term by term:

$$\int f(x) \, dx = C + \sum_{n=0}^{\infty} \frac{a_n (x - c)^{n+1}}{n + 1}.$$

Both $f'(x)$ and $\int f(x) \, dx$ have radii of convergence R but the endpoints $c - R$ and $c + R$ must still be checked individually.

Section 10.9: Taylor and Maclaurin Series

1) Find $f'(x)$ for

$$f(x) = \sum_{n=0}^{\infty} 2^n (x-1)^n.$$

$$f'(x) = \sum_{n=1}^{\infty} 2^n\, n\, (x-1)^{n-1}.$$

2) Find $\int f(x)\, dx$ where

$$f(x) = \sum_{n=0}^{\infty} \frac{(x-3)^n}{n!}.$$

$$\int f(x)\, dx = C + \sum_{n=0}^{\infty} \frac{1}{n!} \frac{(x-3)^{n+1}}{n+1}$$

$$= C + \sum_{n=0}^{\infty} \frac{(x-3)^{n+1}}{(n+1)!}.$$

3) Is $f(x) = \sum_{n=0}^{\infty} \frac{x^n}{4^n}$ continuous at $x = 2$?

Yes. 2 is in $[-4, 4]$, the interval of convergence for $f(x)$.

B. If a function $y = f(x)$ can be expressed as a power series it will be in this form, called the <u>Taylor series of f at c</u>:

$$f(x) = \sum_{n=0}^{\infty} \frac{f^{(n)}(c)}{n!} (x-c)^n$$

$$= f(c) + f'(c)(x-c) + \frac{f''(c)}{2!}(x-c)^2 + \ldots + \frac{f^{(n)}(c)}{n!}(x-c)^n + \ldots$$

In the special case of $c = 0$, this is called the <u>Maclaurin series of f</u>.
To find a Taylor series for a given $y = f(x)$, you must find a formula for
$f^{(n)}(c)$ usually in terms of n. Computing the first few derivatives $f'(c), f''(c),$
$f'''(c), f^{(4)}(c), \dots$ often helps.

For some functions it is easier to find the Taylor series for $f'(x)$ or $\int f(x)\, dx$

then integrate or differentiate term by term to obtain the series for f.

Here are some basic Maclaurin series:

$$e^x = \sum_{n=0}^{\infty} \frac{x^n}{n!} \text{ for all } x. \qquad\qquad \ln(1 + x) = \sum_{n=0}^{\infty} \frac{(-1)^{n+1}}{n} x^n$$

$$\text{for } x \in (-1, 1].$$

$$\sin x = \sum_{n=0}^{\infty} \frac{(-1)^n}{(2n+1)!} x^{2n+1} \text{ for all } x. \qquad \cos x = \sum_{n=0}^{\infty} \frac{(-1)^n}{(2n)!} x^{2n} \text{ for all } x.$$

$$\frac{1}{1-x} = \sum_{n=0}^{\infty} x^n \text{ for } |x| < 1.$$

4) Find the Taylor series for $\dfrac{1}{x}$ at $c = 1$.

   a) directly from the definition.

We must find the general form of $f^{(n)}(1)$:

$$f(x) = x^{-1} \qquad\qquad f(1) = 1$$
$$f'(x) = -x^{-2} \qquad\quad\; f'(1) = -1$$
$$f''(x) = 2x^{-3} \qquad\quad\; f''(1) = 2$$
$$f'''(x) = -6x^{-4} \qquad\quad f'''(1) = -6$$
$$f^{(4)}(x) = 24x^{-5} \qquad\quad f^{(4)}(1) = 24$$

In general, $f^{(n)}(1) = (-1)^n n!$

Thus the Taylor series is $\displaystyle\sum_{n=0}^{\infty} \frac{(-1)^n n!}{n!} (x - 1)^n$

$$= \sum_{n=0}^{\infty} (-1)^n (x-1)^n = \sum_{n=0}^{\infty} (1-x)^n.$$

b) using substitution in a geometric

series $\left(\text{Hint: } \dfrac{1}{x} = \dfrac{1}{1-(1-x)}.\right)$

In the form $\dfrac{1}{1-(1-x)}$, this is the value

of a geometric series with $a = 1, r = 1 - x$.

Thus $\dfrac{1}{x} = \sum_{n=1}^{\infty} 1(1-x)^{n-1} = \sum_{n=0}^{\infty} (1-x)^n.$

5) Find directly the Maclaurin series for
   $f(x) = e^{4x}$.

$f(x) = e^{4x}$ $\qquad$ $f(0) = 1$
$f'(x) = 4\,e^{4x}$ $\qquad$ $f'(0) = 4$
$f''(x) = 16\,e^{4x}$ $\qquad$ $f''(0) = 16$
In general $f^{(n)}(0) = 4^n$.

Thus $e^{4x} = \sum_{n=0}^{\infty} \dfrac{4^n}{n!}\,x^n.$

6) Obtain the Maclaurin series for $\dfrac{x}{1+x^2}$

from the series for $\dfrac{1}{1-x} = \sum_{n=0}^{\infty} x^n.$

$\dfrac{1}{1-x} = 1 + x + x^2 + \ldots + x^n + \ldots$

Substitute $-x^2$ for x:

$\dfrac{1}{1+x^2} = 1 - x^2 + x^4 + \ldots + (-1)^n x^{2n} + \ldots.$

Now multiply by x:

$$\frac{x}{1 + x^2} = x - x^3 + x^5 + \ldots + (-1)^n x^{2n+1} + \ldots$$

$$= \sum_{n=0}^{\infty} (-1)^n x^{2n+1}$$

7) Using $e^x = \sum_{n=0}^{\infty} \frac{x^n}{n!}$ find the Maclaurin

series for $\sinh x$.

$$\left(\text{Remember, } \sinh x = \frac{e^x - e^{-x}}{2}. \right)$$

$$e^x = 1 + x + \frac{x^2}{2!} + \ldots + \frac{x^n}{n!} + \ldots$$

Thus $e^{-x} = 1 - x + \frac{x^2}{2!} + \ldots + \frac{(-1)^n x^n}{n!} + \ldots$

and subtracting term by term (the even terms cancel)

$$e^x - e^{-x} = 2x + \frac{2x^3}{3!} + \ldots + \frac{2x^{2n+1}}{(2n+1)!} + \ldots$$

Thus $\sinh x = x + \frac{x^3}{3!} + \ldots + \frac{x^{2n+1}}{(2n+1)!} + \ldots$

$$= \sum_{n=0}^{\infty} \frac{x^{2n+1}}{(2n+1)!}.$$

8) Find an infinite series expression for

$$\int_0^{1/2} \frac{1}{1 - x^3} \, dx.$$

$$\frac{1}{1 - x} = \sum_{n=0}^{\infty} x^n. \quad \text{Thus} \quad \frac{1}{1 - x^3} = \sum_{n=0}^{\infty} x^{3n}.$$

$$\int \frac{1}{1 - x^3} \, dx = \int \sum_{n=0}^{\infty} x^{3n} \, dx$$

$$= \sum_{n=0}^{\infty} \frac{x^{3n+1}}{3n+1} = F(x).$$

$$\int_0^{1/2} \frac{1}{1-x^3}\,dx = F(\frac{1}{2}) - F(0)$$

$$= \sum_{n=0}^{\infty} \frac{\left(\frac{1}{2}\right)^{3n+1}}{3n+1} - 0$$

$$= \sum_{n=0}^{\infty} \frac{1}{(6n+2)\,8^n}.$$

C. Convergent series may be multiplied and divided like polynomials. Often finding the first few terms of the result is sufficient.

9) Find the first three terms of the Maclaurin series for $e^x \sin x$.

$$e^x = 1 + x + \frac{x^2}{2} + \frac{x^3}{6} + \dots$$

$$\sin x = x - \frac{x^3}{3!} + \frac{x^5}{5!} - \dots$$

Thus $e^x \sin x =$

$$x\left(1 + x + \frac{x^2}{2} + \frac{x^3}{6} + \dots\right)$$

$$-\frac{x^3}{6}\left(1 + x + \frac{x^2}{2} + \frac{x^3}{6} + \dots\right)$$

$$+\frac{x^5}{120}\left(1 + x + \frac{x^2}{2} + \frac{x^3}{6} + \dots\right)$$

$$+\dots$$

$$=\left(x + x^2 + \frac{x^3}{2} + \frac{x^4}{6} + \dots\right)$$

$$-\left(\frac{x^3}{6} + \frac{x^4}{6} + \frac{x^5}{12} + \frac{x^6}{36} + \ldots\right)$$

$$+\left(\frac{x^5}{120} + \frac{x^6}{120} + \frac{x^7}{240} + \ldots\right)$$

$$= x + x^2 + \frac{x^3}{3} + \ldots$$

The Binomial Series

Concepts to Master

Binomial coefficients; Binomial series for $(1 + x)^k$

Summary and Focus Questions

For any real number k and $|x| < 1$, the <u>binomial series for $(1 + x)^k$</u> is the Maclaurin series for $(1 + x)^k$:

$$(1 + x)^k = \sum_{n=0}^{\infty} \binom{k}{n} x^n = 1 + kx + \frac{k(k-1)}{2}x^2 + \ldots + \binom{k}{n}x^n + \ldots$$

Here $\binom{k}{n}$ is the <u>binomial coefficient</u>, which is defined as

$$\binom{k}{n} = \frac{k(k-1) \ldots (k-n+1)}{n!}$$

Also $\binom{k}{0} = 1$ for all k.

The binomial series converges to $(1 + x)^k$ for these cases

Condition on k	Interval of Convergence
$k \leq -1$	$(-1, 1)$
$-1 < k < 0$	$(-1, 1]$
k nonnegative integer	$(-\infty, \infty)$
$k > 0$, but not an integer	$[-1, 1]$

Section 10.10: The Binomial Series

1) Find $\begin{pmatrix} 4/3 \\ 5 \end{pmatrix}$.

$$\begin{pmatrix} 4/3 \\ 5 \end{pmatrix} = \frac{(4/3)(1/3)(-2/3)(-5/3)(-8/3)}{5 \cdot 4 \cdot 3 \cdot 2 \cdot 1}$$

$$= \frac{-8}{3^6} = \frac{-8}{729}.$$

2) True or False:

If k is a nonnegative integer, then the binomial series for $(1 + x)^k$ is a finite sum.

True.

3) Find the binomial series for $\sqrt[3]{1 + x}$.

$$\sum_{n=0}^{\infty} \begin{pmatrix} \frac{1}{3} \\ n \end{pmatrix} x^n.$$

4) Find a power series for $f(x) = \sqrt{4 + x}$.

$$\sqrt{4 + x} = 2\left(1 + \frac{x}{2}\right)^{1/2}$$

$$= 2 \sum_{n=0}^{\infty} \begin{pmatrix} \frac{1}{2} \\ n \end{pmatrix} \left(\frac{x}{2}\right)^n$$

$$= \sum_{n=0}^{\infty} \begin{pmatrix} \frac{1}{2} \\ n \end{pmatrix} \frac{x^n}{2^{n-1}}.$$

5) Find an infinite series for $\sin^{-1}x$.
   (Hint: Start with $(1 + x)^{-1/2}$.)

$$\frac{1}{\sqrt{1+x}} = (1+x)^{-1/2} = \sum_{n=0}^{\infty} \binom{-1/2}{n} x^n.$$

$$\frac{1}{\sqrt{1-x}} = \sum_{n=0}^{\infty} \binom{-1/2}{n} (-x)^n$$

$$= \sum_{n=0}^{\infty} (-1)^n \binom{-1/2}{n} x^n.$$

$$\frac{1}{\sqrt{1-x^2}} = \sum_{n=0}^{\infty} (-1)^n \binom{-1/2}{n} x^{2n}.$$

Now integrate term by term:

$$\sin^{-1} x = \sum_{n=0}^{\infty} \frac{(-1)^n \binom{-1/2}{n} x^{2n+1}}{2n+1}.$$

Approximation by Taylor Polynomials

Concepts to Master

A. Taylor polynomials; Approximating function values

B. Taylor's Formula with Remainder (Lagrange's form); Estimating error in approximations

Summary and Focus Questions

A. If $f^{(n)}(c)$ exists, the <u>Taylor polynomial of degree n for f about c</u> is

$$T_n(x) = f(c) + \frac{f'(c)}{1!}(x-c) + \frac{f''(c)}{2!}(x-c)^2 + \ldots + \frac{f^{(n)}(c)}{n!}(x-c)^n.$$

$T_n(x)$ is simply the nth partial sum of the Taylor series for $f(x)$.

$T_1(x) = f(c) + f'(c)(x-c)$ is the familiar equation for the tangent line to $y = f(x)$ at $x = c$. $T_2(x)$ would be (if there were such a phrase) the "tangent parabola."

Since the first n derivatives of T_n at c are equal to the corresponding first n derivatives of f at c, $T_n(x)$ is an approximation for $f(x)$ if x is near c. Recall that we used $T_1(x)$ for approximations in Section 2.9.

1) Let $f(x) = \sqrt{x}$.

    a) Construct the Taylor polynomial
      of degree 3 for $f(x)$ about $x = 1$.

$$f(x) = x^{1/2},\ f(1) = 1$$

$$f'(x) = \frac{1}{2}x^{-1/2}, f'(1) = \frac{1}{2}$$

Section 10.11: Approximation by Taylor Polynomials

$$f''(x) = -\frac{1}{4}x^{-3/2}, \quad f''(1) = -\frac{1}{4}$$

$$f^{(3)}(x) = \frac{3}{8}x^{-5/2}, \quad f^{(3)}(1) = \frac{3}{8}$$

$$T_3(x) = 1 + \frac{1}{2}(x-1) - \frac{1/4}{2!}(x-1)^2 + \frac{3/8}{3!}(x-1)^3$$

$$= 1 + \frac{1}{2}(x-1) - \frac{1}{8}(x-1)^2 + \frac{1}{16}(x-1)^3$$

b) Approximate $\sqrt{3/2}$ using the Taylor polynomial of degree 3 from part a).

$$T_3\left(\frac{3}{2}\right) = 1 + \frac{1}{2}\left(\frac{1}{2}\right) - \frac{1}{8}\left(\frac{1}{2}\right)^2 + \frac{1}{16}\left(\frac{1}{2}\right)^3$$

$$= \frac{79}{64} \approx 1.2344.$$

$$\left(\text{To 4 decimals } \sqrt{3/2} = 1.2247.\right)$$

B. <u>Taylor's Formula with Remainder (Lagrange's form)</u>:

Let f be a function whose $n + 1$ derivatives exist and are continuous in an interval I containing c. Then for $x \in I$,

$$f(x) = T_n(x) + R_n(x)$$

where $R_n(x) = \dfrac{f^{(n+1)}(z)}{(n+1)!}(x-c)^{n+1}$ for some number z between c and x.

$R_n(x)$ is expressed in <u>Lagrange's form</u> for the remainder and is the <u>error</u> in approximating $f(x)$ with $T_n(x)$.

If f has derivatives of all orders and $\lim\limits_{n \to \infty} R_n(x) = 0$, then $f(x)$ is equal to its Taylor series on its interval of convergence. Thus the approximation $T_n(x)$ often can be made accurate to within any given number ε by selecting n such that $|R_n(x)| < \varepsilon$. This usually requires finding upper bound estimates for $|f^{(n+1)}(z)|$.

2) True or False:

In general, the larger the number n, the closer the Taylor polynomial approximation is to the actual functional value.

True.

3) a) Write Taylor's Formula with Remainder with $n = 3$ for $f(x) = x^{5/2}$ about $c = 4$.

$$f(x) = x^{5/2} \qquad\qquad f(4) = 32$$
$$f'(x) = \frac{5}{2}x^{3/2} \qquad\qquad f'(4) = 20$$
$$f''(x) = \frac{15}{4}x^{1/2} \qquad\qquad f''(4) = \frac{15}{2}$$
$$f'''(x) = \frac{15}{8}x^{-1/2} \qquad f'''(4) = \frac{15}{16}$$

$T_3(x) =$

$$32 + 20(x-4) + \frac{15/2}{2!}(x-4)^2 + \frac{15/16}{3!}(x-4)^3$$

$$= 32 + 20(x-4) + \frac{15}{4}(x-4)^2 + \frac{15}{32}(x-4)^3.$$

Since $f^{(4)}(x) = -\frac{15}{16}x^{-3/2}$,

$$R_3(x) = \frac{-15/16\, z^{-3/2}}{4!}(x-4)^4. \text{ Therefore}$$

$$R_3(x) = \frac{-5}{128\, z^{3/2}}(x-4)^4, \text{ for } z \text{ between } c \text{ and } x.$$

Taylor's Formula is $f(x) = T_3(x) + R_3(x)$, where $T_3(x)$ and $R_3(x)$ are given above.

Section 10.11: Approximation by Taylor Polynomials

b) Estimate the error between
$f(5) = 5^{5/2}$ and $T_3(5)$.

An estimate of the error is

$$|R_3(5)| = \frac{5}{128\, z^{3/2}}(5-4)^4 = \frac{5}{128\, z^{3/2}}.$$

Since $4 < z < 5$, $8 = 4^{3/2} < z^{3/2}$.

Thus $|R_3(5)| < \frac{5}{128(8)} = \frac{5}{1024} \approx .00488.$

c) Compute $T_3(5)$.

$$T_3(5) = 32 + 20(1)^1 + \frac{15}{4}(1)^2 + \frac{5}{32}(1)^3$$

$$= \frac{1789}{32} \approx 55.9063.$$

d) To four decimal places, what is the actual error?

To four decimals $f(5) = 5^{5/2} = 55.9017.$

Thus the error is $55.9063 - 55.9017 = .0046.$

4) a) What degree Maclaurin polynomial is needed to approximate cos .5 accurate to within .00001?

Rephrased, the question asks: For what n is $|R_n(.5)| < .00001$ when $f(x) = \cos x$ and $c = 0$?

$$|R_n(.5)| = \left|\frac{f^{(n+1)}(z)}{(n+1)!}(.5-0)^{n+1}\right|$$

$$= \frac{\left|f^{(n+1)}(z)\right|(.5)^{n+1}}{(n+1)!} \le \frac{(.5)^{n+1}}{(n+1)!}$$

since $\left|f^{(n+1)}(z)\right| \le 1$.

Now $\frac{(.5)^{n+1}}{(n+1)!} \ge .00001$ for $n = 1, 2, 3, 4$.

However, for $n = 5$, $\frac{(.5)^{n+1}}{(n+1)!} = .000022 < .00001$.

Thus the degree should be at least 5.

b) Use your answer to part a) to estimate cos .5 to within .00001.

For $f(x) = \cos x$, $f(0) = 1$,
$f'(0) = 0$, $f''(0) = -1$,
$f^{(3)}(0) = 0$, $f^{(4)}(0) = 1$, and
$f^{(5)}(0) = 0$.
Thus $T_5(x) = 1 - \dfrac{x}{2!} + \dfrac{x}{4!}$ and $T_5(.5) \approx .87760$.

Review and Preview 1

_____ **1.** True, False:

Any number that can be expressed as a decimal is a real number.

_____ **2.** True, False:

The intervals [5, 9) and [8, 13] have no points in common (do not intersect.)

_____ **3.** The solution to $3x - 5 < 14x + 17$ is

    a) $x > 2$
    b) $x < 2$
    c) $x > -2$
    d) $x < -2$

_____ **4.** The solution set for $x^4 - x^2 \geq 0$ is

    a) $[0, \infty)$
    b) $(-\infty, 0]$
    c) $(-\infty, -1] \cup [1, \infty]$
    d) $[-1, 1]$

_____ **5.** Sometimes, Always, or Never: $|x - 3| = 3 - x$.

_____ **6.** The solution set for $|5x - 1| \leq 11$ is

    a) $[-2, 12/5]$
    b) $[-12/5, 2]$
    c) $[-12/5, -2]$
    d) $[2, 12/5]$

Review and Preview 2

_____ **1.** The distance between (6, 5) and (10, 2) is

       a) $\sqrt{7}$
       b) 25
       c) $\sqrt{41}$
       d) 5

_____ **2.** True, False: $y = 3x + 12$ is perpendicular to $y = \frac{1}{3}x - \frac{1}{12}$.

_____ **3.** The slope of the line through (5, 10) and (3, 5) is

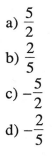

       a) $\frac{5}{2}$

       b) $\frac{2}{5}$

       c) $-\frac{5}{2}$

       d) $-\frac{2}{5}$

_____ **4.** The equation of the line through (7, 0) with slope 2 is

       a) $y = 2x + 7$
       b) $y = 2(x - 7)$
       c) $y = 7x + 2$
       d) $y = 7(x - 2)$

_____ **5.** The point of intersection of $2x + 3y = 7$ with $3x + y = -5$ is

       a) (2, 2)
       b) (1, 2)
       c) (3, 1)
       d) (2, 1)

Review and Preview 3

_____ **1.** The graph of $\dfrac{x^2}{10} - \dfrac{y^2}{15} = 1$ is a(n):

     a) ellipse
     b) parabola opening upward
     c) parabola opening downward
     d) hyperbola opening upward and downward
     e) hyperbola opening to the left and right

_____ **2.** True, False:

The graph of $y = (x - 5)^2$ is the graph of $y = x^2$ which has been shifted 5 units

     a) upward
     b) downward
     c) to the left
     d) to the right

_____ **3.** The center of the circle $x^2 - 6x + y^2 + 4y = 12$ is

     a) $(3, 2)$
     b) $(3, -2)$
     c) $(-3, 2)$
     d) $(-3, -2)$

_____ **4.** The radius of the circle $x^2 + 6x + y^2 + 4y = 12$ is

     a) 12
     b) $\sqrt{23}$
     c) 5
     d) 25

_____ **1.** True, False: $x^2 + 6x + 2y = 1$ defines y as a function of x.

_____ **2.** True, False: In $V = \dfrac{4}{3}\pi r^3$, r is the dependent variable.

_____ **3.** The implied domain of $f(x) = \dfrac{1}{\sqrt{1-x}}$ is

      a) $(1, \infty)$
      b) $(-\infty, 1)$
      c) $x \neq 1$
      d) $(-1, 1)$

_____ **4.** True, False: The graph below is the graph of an even function.

_____ **5.** Which graph best represents $y = x + \dfrac{1}{x}$?

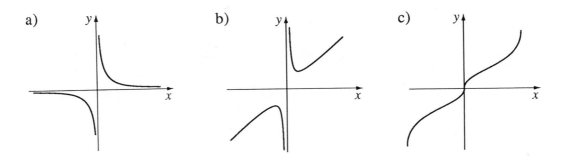

a)   b)   c)

_____ **1.** For $f(x) = 3x + 1$ and $g(x) = x$, $\dfrac{f}{g}(x) =$

a) $\dfrac{3 + x}{x}$

b) $\dfrac{x}{3x + 1}$

c) $\dfrac{3x + 1}{x}$

d) $3x + \dfrac{1}{x}$

_____ **2.** For $f(x) = \sqrt{x^2 + x}$, we may write $f(x) = h \circ g(x)$, where

a) $h(x) = \sqrt{x}$ and $g(x) = x^2 + x$
b) $h(x) = x^2 + x$ and $g(x) = \sqrt{x}$
c) $h(x) = x^2$ and $g(x) = \sqrt{x}$
d) $h(x) = x^2 + x$ and $g(x) = x^2$

_____ **3.** Let $f(x) = 2 + \sqrt{x}$ and $g(x) = x + 3$. Then $g \circ f(x) =$

a) $2 + \sqrt{x} + 3$
b) $2 + \sqrt{x + 3}$
c) $3 + \sqrt{x + 2}$
d) $(x + 3)\left(2 + \sqrt{x}\right)$

_____ **4.** Sometimes, Always, or Never: $f \circ g(x) = g \circ f(x)$.

_____ **5.** For $f(x) = \cos(x^2 + 1)$ we may write $f(x) = h \circ g(x)$, where

a) $h(x) = \cos x^2$ and $g(x) = x + 1$
b) $h(x) = \cos x$ and $g(x) = x^2 + 1$
c) $h(x) = x^2 + 1$ and $g(x) = \cos x$
d) $h(x) = x^2$ and $g(x) = \cos(x + 1)$

Review and Preview 6

_____ **1.** True, False:

$$f(x) = \frac{x^2 + \sqrt{x}}{2x + 1}$$ is a rational function.

_____ **2.** True, False:

The degree of $f(x) = 4x^3 + 7x^6 + 1$ is 7.

_____ **3.** The graph of $y = f(3x)$ is obtained from the graph of $y = f(x)$ by

    a) stretching vertically by a factor of 3.
    b) compressing vertically by a factor of 3.
    c) stretching horizontally by a factor of 3.
    d) compressing horizontally by a factor of 3.

_____ **4.** True, False:

Given the graph of $y = f(x)$ on the left below, the graph on the right below is $y = -f(x)$.

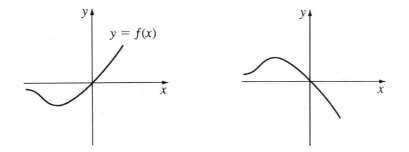

_____ **1.** For $f(x) = 5x^2 + 1$ the slope of the secant line between the points corresponding to $x = 3$ and $x = 4$ is

      a) 7
      b) 35
      c) 81
      d) 1

_____ **2.** A guess for the slope of the tangent line to $f(x) = 5x^2 + 1$ at $P(3, 26)$ based on the chart to the right is

x	$f(x)$
4	81
3.1	49.05
3.01	46.3005
3.001	46.030005

      a) 26
      b) 30
      c) 35
      d) 46

_____ **3.** True, False:

The slope of a tangent line may be interpreted as average velocity.

Section 1.2

_____ **1.** In the graph at the right, $\lim_{x \to 2} f(x) =$

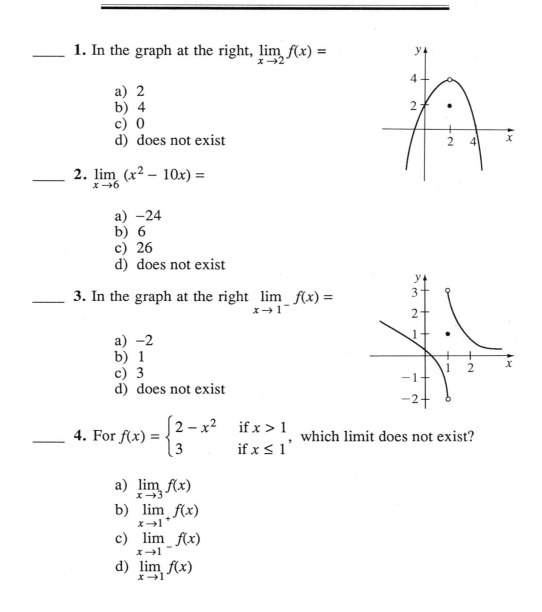

a) 2
b) 4
c) 0
d) does not exist

_____ **2.** $\lim_{x \to 6} (x^2 - 10x) =$

a) −24
b) 6
c) 26
d) does not exist

_____ **3.** In the graph at the right $\lim_{x \to 1^-} f(x) =$

a) −2
b) 1
c) 3
d) does not exist

_____ **4.** For $f(x) = \begin{cases} 2 - x^2 & \text{if } x > 1 \\ 3 & \text{if } x \le 1 \end{cases}$, which limit does not exist?

a) $\lim_{x \to 3} f(x)$
b) $\lim_{x \to 1^+} f(x)$
c) $\lim_{x \to 1^-} f(x)$
d) $\lim_{x \to 1} f(x)$

_____ **1.** True, False:

If $h(x) = g(x)$ for all $x \neq a$ and $\lim_{x \to a} h(x) = L$, then $\lim_{x \to a} g(x) = L$.

_____ **2.** $\lim_{x \to 3} \dfrac{7x + 6}{5x - 12} =$

    a) $-\dfrac{1}{2}$

    b) $\dfrac{7}{5}$

    c) 9

    d) 4

_____ **3.** Sometimes, Always, or Never: $\quad \lim_{x \to a} \dfrac{f(x)}{g(x)} = \dfrac{\lim\limits_{x \to a} f(x)}{\lim\limits_{x \to a} g(x)}$.

_____ **4.** $\lim_{x \to 2} \dfrac{x^2(x - 1)}{1 - x} =$

    a) 1
    b) −1
    c) 0
    d) does not exist

_____ **5.** $\lim_{x \to 1} \dfrac{x^3 - x^2}{1 - x} =$

    a) 1
    b) −1
    c) 0
    d) does not exist

_____ **6.** If $2x + 2 \leq f(x) \leq x^2 + 3$, then $\lim_{x \to 1} f(x) =$
    a) 1
    b) 4
    c) does not exist
    d) exists but cannot be determined from the information given

Section 1.4

_____ **1.** True, False:

$|f(x) - L| < \varepsilon$, means that the distance between $f(x)$ and L is less than ε.

_____ **2.** For a given $\varepsilon > 0$, the corresponding δ in the definition of
$\lim\limits_{x \to 6} 3x + 5 = 23$ is

    a) $\dfrac{\varepsilon}{6}$

    b) $\dfrac{\varepsilon}{3}$

    c) $\dfrac{\varepsilon}{5}$

    d) $\dfrac{\varepsilon}{23}$

_____ **1.** Sometimes, Always, or Never:

If $\lim_{x \to a} f(x)$ and $f(a)$ both exist, then f is continuous at a.

_____ **2.** $f(x) = \dfrac{x + 2}{x(x - 5)}$ is continuous for all x except

      a) $x = -2$
      b) $x = -2, 0, 5$
      c) $x = 0, 5$
      d) f is continuous for all x.

_____ **3.** True, False:

$f(x) = \dfrac{x^2(x + 1)}{x}$ is continuous for all real numbers.

      a) $[0, \infty)$
      b) $(-\infty, 0]$
      c) $(-\infty, -1] \cup [1, \infty]$
      d) $[-1, 1]$

_____ **4.** For the continuous function at the right, how many points c does the Intermediate Value Theorem guarantee must be between a and b and have $f(c) = N$?

      a) 1
      b) 3
      c) 5
      d) none

_____ **5.** True, False:

Let $f(x) = x^2$. The Intermediate Value Theorem guarantees there is $w \in (-1, 5)$ for which $f(w) = 0$.

_____ **1.** Sometimes, Always, or Never: $\lim\limits_{x\to\infty} f(x) = \lim\limits_{x\to -\infty} f(x)$.

_____ **2.** $\lim\limits_{x\to\infty} \dfrac{\sqrt{x^2+1}}{3x+2} =$

     a) 0

     b) 1

     c) $\dfrac{1}{3}$

     d) does not exist

_____ **3.** True, False:

If the limits exist, then $\lim\limits_{x\to\infty} (f(x) + g(x)) = \lim\limits_{x\to\infty} f(x) + \lim\limits_{x\to\infty} g(x)$.

_____ **4.** The horizontal asymptote(s) for $f(x) = \dfrac{|x|}{x+1}$ is (are)

     a) $y = 0$

     b) $y = 1$

     c) $y = -1$

     d) $y = 1$ and $y = -1$.

_____ **5.** True, False:

$$\lim_{x\to\infty} \frac{6x^3 + 8x^2 - 9x + 7}{5x^3 - 4x + 30} = \lim_{x\to\infty} \frac{6x^3}{5x^3}.$$

Section 1.7

_____ **1.** $\lim\limits_{x \to 0^+} \dfrac{x+1}{x(x+2)} =$

     a) 0

     b) $\dfrac{1}{2}$

     c) ∞

     d) does not exist

_____ **2.** $\lim\limits_{x \to \infty} \dfrac{x^3+1}{x^2-1} =$

     a) ∞

     b) 1

     c) $\dfrac{3}{2}$

     d) does not exist

_____ **3.** Sometimes, Always, or Never:

If $\lim\limits_{x \to a} f(x) = \infty$ and $\lim\limits_{x \to a} g(x) = \infty$, then $\lim\limits_{x \to a} \dfrac{f(x)}{g(x)} = 1$.

_____ **4.** The vertical asymptote(s) of $f(x) = \dfrac{x(x-3)}{(x-3)(x-5)}$ is (are)

     a) $x = 3$ and $x = 5$

     b) $x = 3$

     c) $x = 5$

     d) $x = 0$ and $x = 5$

Section 1.8

_____ **1.** True, False:

$$\lim_{x \to a} \frac{f(x) - f(a)}{x - a} = \lim_{k \to 0} \frac{f(a + k) - f(a)}{k}.$$

_____ **2.** True, False:

$\dfrac{f(x) - f(a)}{x - a}$ may be interpreted as the instantaneous velocity of a particle at time a.

_____ **3.** The slope of the tangent line to $y = x^3$ at $x = 2$ is

    a) 18
    b) 12
    c) 6
    d) 0

_____ **4.** True, False:

Slope of a tangent line, instantaneous velocity, and instantaneous rate of change of a function are each an interpretation of the same limit concept.

_____ **5.** Given the table of function values at the right, which of the following is the best estimate for the instantaneous rate of change of $y = f(x)$ at $x = 4$?

x	$f(x)$
6	17
5	13.3
4.5	11.7
4.1	10.32
4.05	10.16
4.01	10.031
4	10

    a) 17
    b) 10
    c) 3
    d) 0

Section 2.1

____ **1.** True, False: $\quad f'(x) = \lim\limits_{x \to 0} \dfrac{f(x + h) - f(x)}{h}$

____ **2.** True, False: $\quad f'(x) = \lim\limits_{x \to a} \dfrac{f(x) - f(a)}{x - a}$

____ **3.** For $f(x) = 10x^2, f'(3) =$

      a) 10
      b) 20
      c) 30
      d) 60

____ **4.** The instantaneous rate of change of $y = x^3 + 3x$ at the point corresponding to $x = 2$ is

      a) 10
      b) 15
      c) 2
      d) 60

____ **5.** $\lim\limits_{h \to 0} \dfrac{\sqrt{4 + h} - 2}{h}$ is the derivative of

      a) $f(x) = \sqrt{4 + x}$ at $x = 2$
      b) $f(x) = \sqrt{x}$ at $x = 4$
      c) $f(x) = \sqrt{x}$ at $x = 2$
      d) $f(x) = \dfrac{\sqrt{4 + x} - 2}{h}$ at $x = 0$

____ **6.** True, False: $\quad \dfrac{dm}{dv}$ stands for the derivative of the function m with respect to the variable v.

_____ **1.** For $f(x) = 6x^4$, $f'(x) =$

    a) $24x$
    b) $6x^3$
    c) $10x^3$
    d) $24x^3$

_____ **2.** For $y = \dfrac{1}{2} g t^2$, $\dfrac{dy}{dt} =$

    a) $\dfrac{1}{4} g^2$

    b) $\dfrac{1}{2} g$

    c) $g t$
    d) $g t^2$

_____ **3.** True, False: $(f(x)\, g(x))' = f'(x)\, g'(x)$.

_____ **4.** For $f(x) = \dfrac{x}{x + 1}$, $f'(x) =$

    a) 1

    b) $\dfrac{x^2}{(x + 1)^2}$

    c) $\dfrac{-1}{(x + 1)^2}$

    d) $\dfrac{1}{(x + 1)^2}$

_____ **5.** For $f(x) = 5 + g(x)$, $f'(x) =$

    a) $5\, g'(x)$
    b) $5 + g'(x)$
    c) $0 \cdot g'(x)$
    d) $g'(x)$

_____ **1.** True, False: $f'(t)$ is used to measure the average rate of change of f with respect to t.

_____ **2.** Air is being pumped from a chamber so that after t seconds the pressure in the chamber is $3 + \dfrac{1}{x^2 + 5}$ pounds/in². The rate of change of the pressure at $t = 1$ second is

    a) $-\dfrac{1}{3}$ pounds/in²/sec

    b) $3\dfrac{1}{6}$ pounds/in²/sec

    c) $3\dfrac{1}{2}$ pounds/in²/sec

    d) $-2\dfrac{5}{6}$ pounds/in²/sec

_____ **3.** A bacteria population grows such that after t hours it is $1000 + 10t + t^3$. The growth rate after 3 hours is

    a) 13 bacteria/hour
    b) 37 bacteria/hour
    c) 10 bacteria/hour
    d) 1057 bacteria/hour

_____ **1.** $\lim\limits_{x \to 0} \dfrac{2 \sin x}{3x} =$

      a) 1

      b) $\dfrac{2}{3}$

      c) $\dfrac{3}{2}$

      d) 0

_____ **2.** If $\lim\limits_{x \to 0} \dfrac{\sin x}{x} = 1$, the angle x must be measured in

      a) radians

      b) degrees

      c) it does not matter

_____ **3.** For $y = x \sin x$, $y' =$

      a) $x \cos x$

      b) $x \cos x + 1$

      c) $\cos x$

      d) $x \cos x + \sin x$

_____ **4.** For $y = 3x + \tan x$, $y' =$

      a) $3 + \sec x$

      b) $3x \sec^2 x + 3 \tan x$

      c) $3 + \sec^2 x$

      d) $3x + \sec^2 x$

_____ **1.** True, False: If $y = f(g(x))$ then $y' = f'(g'(x))$.

_____ **2.** For $y = (x^2 + 1)^2$, $y' =$

    a) $2(x^2 + 1)$
    b) $4(x^2 + 1)$
    c) $4x(x^2 + 1)$
    d) $2x(x^2 + 1)$

_____ **3.** For $y = \sin 6x$, $\dfrac{dy}{dx} =$

    a) $\cos 6x$
    b) $6 \cos x$
    c) $\cos 6$
    d) $6 \cos 6x$

_____ **4.** For $y = \dfrac{4}{\sqrt{3 + 2x}}$, $y' =$

    a) $-4(3 + 2x)^{-3/2}$
    b) $4(3 + 2x)^{-3/2}$
    c) $-2(3 + 2x)^{-1/2}$
    d) $2(3 + 2x)^{-1/2}$

_____ **1.** True, False:

$y = \sqrt[3]{4 - \sqrt{x^2 + 1}}$ defines y implicitly as a function of x.

_____ **2.** If y is defined as a function of x by $y^3 = x^2$, then $y' =$

a) $\dfrac{2x}{y^3}$

b) $\dfrac{2x}{3y^2}$

c) $\dfrac{x^2}{3y^2}$

d) $\dfrac{x^2}{3y}$

_____ **3.** The slope the line tangent to the circle $x^2 + y^2 = 100$ at the point $(-6, 8)$ is

a) $\dfrac{3}{4}$

b) $-\dfrac{3}{4}$

c) $\dfrac{4}{3}$

d) $-\dfrac{4}{3}$

Section 2.7

_____ **1.** True, False:

$y^{(6)}$ is the derivative of $y^{(5)}$.

_____ **2.** For $y = \dfrac{5}{x^3}$, $y'' =$

    a) $\dfrac{5}{6x}$

    b) $\dfrac{5}{12x}$

    c) $-\dfrac{60}{x^5}$

    d) $60x^{-5}$

_____ **3.** For $y = x^n$ (n a positive integer), $y^{(n)} =$

    a) 0
    b) $n!$
    c) $n\,x^{n-1}$
    d) $n!\,x$

_____ **4.** For a particle whose position at time t (seconds) is
$t^3 - 3t^2 + 10t + 1$ meters, its acceleration at time $t = 3$ is

    a) 19 m/sec^2
    b) 18 m/sec^2
    c) 0 m/sec^2
    d) -5 m/sec^2

_____ **5.** For $y = \cos x$, $y^{(14)} =$

    a) $\cos x$
    b) $\sin x$
    c) $-\cos x$
    d) $-\sin x$

_____ **1.** A right triangle has one leg with constant length 8 cm. The length of the other leg is decreasing at 3 cm/sec. The rate of change of the hypotenuse when the variable leg is 6 cm is

a) $-\dfrac{9}{5}$ cm/sec

b) 3 cm/sec

c) $-\dfrac{5}{3}$ cm/sec

d) $\dfrac{5}{3}$ cm/sec

_____ **2.** How fast is the angle between the hands of a clock increasing at 3:30 pm?

a) 2π radians/hr

b) $\dfrac{\pi}{6}$ radians/hr

c) $\dfrac{11\pi}{6}$ radians/hr

d) $\dfrac{5\pi}{6}$ radians/hr

_____ **1.** For $f(x) = x^3 + 7x$, $df =$

     a) $(x^3 + 7x)\, dx$
     b) $(3x^2 + 7)\, dx$
     c) $3x^2 + 7$
     d) $x^3 + 7x + dx$

_____ **2.** The best approximation of Δy using differentials for $y = x^2 - 4x$ at $x = 6$ when $\Delta x = dx = .3$ is

     a) .36
     b) .24
     c) .20
     d) .30

_____ **3.** The linearization of $f(x) = \dfrac{1}{\sqrt{x}}$ at $x = 4$ is

     a) $L(x) = 2 + \dfrac{1}{\sqrt{x}}\,(x - 2)$

     b) $L(x) = 4 + \dfrac{1}{2}\,(x - 2)$

     c) $L(x) = 2 + \dfrac{1}{2}\,(x - 4)$

     d) $L(x) = 2 - \dfrac{1}{16}\,(x - 4)$

_____ **4.** Using differentials, an approximation to $(3.04)^3$ is

     a) 28.094464
     b) 28.08
     c) 28.04
     d) 28

_____ **1.** Sometimes, Always, or Never:

The value of x_2 in Newton's Method will be a closer approximation to the root than x_1.

_____ **2.** With an initial estimate of -1, use Newton's Method once to estimate a zero of $f(x) = 5x^4 + 10x + 3$.

        a) $-\dfrac{16}{15}$

        b) $-\dfrac{14}{15}$

        c) $-\dfrac{7}{5}$

        d) $-\dfrac{3}{5}$

_____ **3.** Suppose in an estimation of a root of $f(x) = 0$ we make an initial guess x_1, so that $f'(x_1) = 0$. Which of the following will be true?

        a) x_1 is a root of $f(x) = 0$.
        b) Newton's Method will produce a sequence $x_1, x_2, \ldots$ that does approximate a zero.
        c) $x_2 = x_1$.
        d) Newton's Method cannot be used.

_____ **1.** Sometimes, Always, or Never: $\lim\limits_{x \to -\infty} a^x = 0.$

_____ **2.** Which is the best graph of $y = 1.5^x$? The graph of $y = 2^x$ is shown as a dashed curve for comparison.

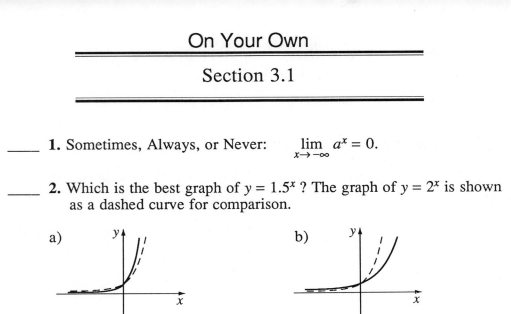

a) b)

c) d)

_____ **3.** $\lim\limits_{x \to -\infty} \left(\dfrac{1}{2}\right)^{x^2} =$

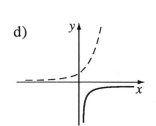

     a) 0
     b) 1
     c) ∞
     d) does not exist

_____ **4.** $\lim\limits_{x \to \infty} 2^{\frac{x}{x+1}} =$

     a) 0
     b) 1
     c) 2
     d) ∞

Section 3.2

_____ **1.** Sometimes, Always, or Never: $(a^x)' = a^x$.

_____ **2.** If $f(x) = 2e^{3x}$ then $f'(x) =$

      a) $2e^{3x}$
      b) $2x\,e^{3x}$
      c) $6e^{3x}$
      d) $6x\,e^{3x}$

_____ **3.** $\lim\limits_{x \to \infty} e^{x/2} =$

      a) e
      b) $\dfrac{1}{e}$
      c) 0
      d) ∞

_____ **4.** If $f(x) = e^{-x}$, then $f^{(7)}(x) =$

      a) e^{-x}
      b) $-e^{-x}$
      c) $7e^{x}$
      d) $-7e^{-x}$

Section 3.3

_____ **1.** A function f is one-to-one means

    a) $x_1 = x_2$, then $f(x_1) = f(x_2)$
    b) $x_1 \neq x_2$, then $f(x_1) = f(x_2)$
    c) $x_1 \neq x_2$, then $f(x_1) \neq f(x_2)$
    d) if $f(x_1) \neq f(x_2)$, then $x_1 \neq x_2$

_____ **2.** If $f(x) = \sqrt[3]{x + 3}$, then $f^{-1}(x) =$

    a) $\dfrac{1}{\sqrt[3]{x + 3}}$
    b) $x^3 + 3$
    c) $(x + 3)^3$
    d) $x^3 - 3$

_____ **3.** Sometimes, Always, or Never:

If f is one-to-one and (a, b) is on the graph of $y = f(x)$ then (b, a) is on the graph of $y = f^{-1}(x)$.

_____ **4.** If f is one-to-one, $f(4) = 7$ and $f'(4) = \dfrac{1}{2}$ then $(f^{-1})'(7) =$

    a) $\dfrac{1}{7}$

    b) $\dfrac{1}{4}$

    c) $\dfrac{1}{2}$

    d) 2

_____ **1.** $\log_{27} 9 =$

    a) $\dfrac{1}{3}$

    b) $\dfrac{2}{3}$

    c) $\dfrac{3}{2}$

    d) $\dfrac{1}{2}$

_____ **2.** True, False:

$\ln(a + b) = \ln a + \ln b.$

_____ **3.** Simplified, $\log_3 9x^3$ is

    a) $2 + 3 \log_3 x$
    b) $6 \log_3 x$
    c) $9 \log_3 x$
    d) $9 + \log_3 x$

_____ **4.** Solve for x: $e^{2x-1} = 10.$

    a) $\dfrac{1}{2}(1 + 10^e)$
    b) $3 + \ln 10$
    c) $2 + \ln 10$
    d) $\ln \sqrt{10e}$

_____ **5.** Solve for x: $\ln(e + x) = 1.$

    a) 0
    b) 1
    c) $-e$
    d) $e^e - e$

_____ **1.** For $y = \ln(3x^2 + 1)$, $y' =$

 a) $\dfrac{1}{3x^2 + 1}$

 b) $\dfrac{6}{x}$

 c) $\dfrac{6x}{3x^2 + 1}$

 d) $\dfrac{1}{6x}$

_____ **2.** For $y = 3^x$, $y' =$

 a) $3^x \log_3 e$

 b) $3^x \ln 3$

 c) $3^x / \log_3 e$

 d) $3^x / \ln 3$

_____ **3.** By logarithmic differentiation, if $y = \dfrac{\sqrt{x}}{x + 1}$, then $y' =$

 a) $\dfrac{\sqrt{x}}{x+1}\left(\dfrac{1}{\sqrt{x}} - \dfrac{1}{x+1}\right)$　　　c) $\dfrac{\sqrt{x}}{x+1}\left(\dfrac{2}{x} - \dfrac{1}{x+1}\right)$

 b) $\dfrac{\sqrt{x}}{x+1}\left(\dfrac{1}{x} - \dfrac{1}{x+1}\right)$　　　d) $\dfrac{\sqrt{x}}{x+1}\left(\dfrac{1}{2x} - \dfrac{1}{x+1}\right)$

_____ **4.** $\displaystyle \lim_{x \to 0^+} (1 + x)^{-1/x} =$

 a) e
 b) $-e$

 c) $\dfrac{1}{e}$

 d) $-\dfrac{1}{e}$

_____ **1.** The general solution to $\dfrac{dy}{dt} = k\,y$ is

    a) $y(t) = y(0)\,e^{kt}$
    b) $y(t) = y(k)\,e^{t}$
    c) $y(t) = y(t)\,e^{k}$
    d) $y(t) = e^{y(0)kt}$

_____ **2.** A bacteria culture starts with 50 organisms and after 2 hours there are 100. How many will there be after 5 hours?

    a) $50 \ln 5$
    b) $50\,e^{5}$
    c) $200\sqrt{2}$
    d) 300

_____ **1.** The range of $f(x) = \cos^{-1}x$ is

    a) $[-\pi/2, \pi/2]$
    b) $[0, \pi]$
    c) $[0, \pi/2] \cup [\pi, 3\pi/2]$
    d) all reals

_____ **2.** $\tan^{-1}\left(-\sqrt{3}\right) =$

    a) $\pi/6$
    b) $-\pi/6$
    c) $\pi/3$
    d) $-\pi/3$

_____ **3.** For $f(x) = \cos^{-1}(2x), f'(x) =$

    a) $\dfrac{-1}{\sqrt{1 - 4x^2}}$

    b) $\dfrac{-2}{\sqrt{1 - 4x^2}}$

    c) $\dfrac{-4}{\sqrt{1 - 4x^2}}$

    d) $\dfrac{-8}{\sqrt{1 - 4x^2}}$

Section 3.8

_____ **1.** True, False: $\sinh^2 x + \cosh^2 x = 1$.

_____ **2.** $\cosh 0 =$

    a) 0

    b) 1

    c) $\dfrac{e^2}{2}$

    d) is not defined

_____ **3.** For $f(x) = \sinh x$, $f''(x) =$

    a) $\cosh^2 x$

    b) $\sinh^2 x$

    c) $\tanh x$

    d) $\sinh x$

_____ **4.** For $f(x) = \tanh^{-1} 2x$, $f'(x) =$

    a) $2(\text{sech}^{-1} 2x)^2$

    b) $2(\text{sech}^2 2x)^{-1}$

    c) $\dfrac{2}{1 - 4x^2}$

    d) $\dfrac{1}{1 - 4x^2}$

____ **1.** $\displaystyle\lim_{x \to 5} \frac{x^2 - 25}{x^2 - 9x + 20} =$

    a) 1
    b) 10
    c) ∞
    d) does not exist

____ **2.** $\displaystyle\lim_{h \to 0^+} \sqrt[3]{x}\ \ln x =$

    a) 0
    b) 1
    c) $-\infty$
    d) does not exist

____ **3.** $\displaystyle\lim_{x \to \infty} (\ln x)^{1/x} =$

    a) 0
    b) 1
    c) e
    d) ∞

Section 4.1

_____ **1.** True, False:

If c is a critical number, then $f'(c) = 0$.

_____ **2.** True, False:

If $f'(c) = 0$, then c is a critical number.

_____ **3.** The critical numbers of $f(x) = 3x^4 + 20x^3 - 36x^2$ are

    a) 0, 1, 6
    b) 0, −1, 6
    c) 0, 1, −6
    d) 0, −1, −6

_____ **4.** True, False:

The absolute extrema of a continuous function on a closed interval always exist.

_____ **5.** The graph at the right has a local maximum at $x =$

    a) 1, 3, and 5
    b) 2
    c) 4
    d) 2 and 4

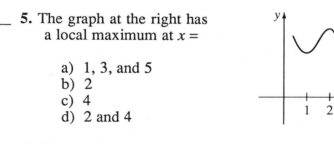

_____ **6.** The absolute maximum value of $f(x) = 6x - x^2$ on $[1, 7]$ is

    a) 3
    b) 9
    c) 7
    d) −7

_____ **1.** Which of the following is not a hypothesis of Rolle's Theorem?

    a) f is continuous on $[a, b]$.
    b) $f(a) = f(b)$.
    c) $f'(x) \geq 0$ for all $x \in (a, b)$.
    d) f is differentiable on (a, b).

_____ **2.** Yes, No:

Do all the hypotheses for the Mean Value Theorem hold for
$f(x) = 1 - |x|$ on $[-1, 2]$?

_____ **3.** A number c that satisfies the Mean Value Theorem for $f(x) = x^3$ on $[1, 4]$ is

    a) $\sqrt{7}$
    b) $\sqrt{21}$
    c) $\sqrt{63}$
    d) 63

_____ **4.** True, False:

If $f'(x) = g'(x)$ for all x then $f(x) = g(x)$.

Section 4.3

_____ **1.** If 4 is a critical number for f and $f'(x) < 0$ for $x < 4$ and $f'(x) > 0$ for $x > 4$ then

      a) f has a local maximum at 4.
      b) f has a local minimum at 4.
      c) 4 is either a local maximum or minimum, but not enough information is given to decide which.
      d) 4 is neither a local maximum nor local minimum.

_____ **2.** True, False:

$f(x) = 10x - x^2$ is monotonic on (4, 8).

_____ **3.** True, False:

$f(x) = 4x^2 - 8x + 1$ is increasing on (0, 1).

_____ **4.** $f(x) = 12x - x^3$ has a local maximum at

      a) $x = 0$
      b) $x = 2$
      c) $x = -2$
      d) f does not have a local maximum.

Section 4.4

_____ **1.** True, False:

If $f''(c) = 0$, then c is a point of inflection for f.

_____ **2.** $f(x) = (x^2 - 3)^2$ has points of inflection at

   a) $0, \sqrt{3}, -\sqrt{3}$
   b) $-1, 1$
   c) $0, -1, 1$
   d) $\sqrt{3}, -\sqrt{3}$

_____ **3.** $f(x) = x^3 - 2x^2 + x - 1$ has a local maximum at

   a) 0
   b) 1
   c) $\dfrac{1}{3}$
   d) f has no local maxima.

_____ **1.** True, False:

The domain is all x for which $f(x)$ is derived.

_____ **2.** True, False:

The x-intercepts are the values of x for which $f(x) = 0$.

_____ **3.** True, False:

$f(x)$ is symmetric about the origin if $f(x) = -f(x)$.

_____ **4.** True, False:

If $f'(x) > 0$ for all x in an interval I then f is increasing on I.

_____ **5.** True, False:

If $f'(c) = 0$ and $f'(c) < 0$, then c is a local minimum.

_____ **6.** True, False:

If $f''(x) > 0$ for $x < c$ and $f''(x) < 0$ for $x > c$, then c is a point of inflection.

_____ **7.** Which is the graph of $f(x) = x + \dfrac{1}{x}$?

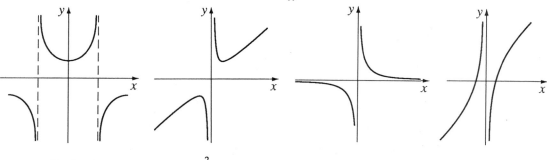

_____ **8.** A sketch of $y = e^{-x^2}$ looks like:

_____ **1.** For two nonnegative numbers, twice the first plus the second is 12. What is the maximum product of two such numbers?

    a) 12
    b) 18
    c) 36
    d) There is no such maximum.

_____ **2.** A carpenter has a 10 ft long board to mark off a triangular area on the floor in the corner of a room. See the figure below. What is the function for the triangular area in terms of x used to determine the maximum such area?

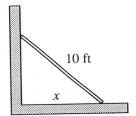

    a) $A = \dfrac{1}{2}x^2$

    b) $A = \dfrac{1}{2}(x^2 + 10)$

    c) $A = \dfrac{1}{2} x \sqrt{100 - x^2}$

    d) $A = \dfrac{1}{2} x \sqrt{100 - 2x^2}$

_____ **1.** True, False:

Average cost is the derivative of total cost.

_____ **2.** True, False:

Marginal revenue is the derivative of total revenue.

_____ **3.** Average cost is minimum when

      a) average cost equals total cost
      b) average cost equals marginal cost
      c) marginal cost equals total cost
      d) marginal cost equals zero

_____ **4.** Profit is maximum when

      a) total revenue equals total cost
      b) marginal revenue equals marginal cost
      c) marginal revenue or marginal cost is zero
      d) marginal revenue is maximum

_____ **5.** Suppose the price (demand function) for an item is $p = 16 - .01x$, where x is the number of items. If the cost function is $C(x) = 1000 + 10x + .02x^2$, how many items should be made and sold to maximize profits?

      a) 50
      b) 100
      c) 200
      d) 250

_____ **1.** True, False:

If $h'(x) = k(x)$, then $k(x)$ is an antiderivative of $h(x)$.

_____ **2.** Find $f(x)$ is $f'(x) = 10x^2 + \cos x$

   a) $20x - \sin x + C$
   b) $20x - \cos x + C$
   c) $\dfrac{10}{3}x - \cos x + C$
   d) $\dfrac{10}{3}x^3 - \sin x + C$

_____ **3.** Find $f(x)$ if $f'(x) = \dfrac{1}{x^2}$ and $f(2) = 0$.

   a) $-\dfrac{1}{x} + \dfrac{1}{2}$
   b) $-\dfrac{1}{x} - \dfrac{1}{2}$
   c) $-\dfrac{3}{x^3} + \dfrac{3}{8}$
   d) $-\dfrac{3}{x^3} - \dfrac{3}{8}$

_____ **1.** $\displaystyle\sum_{i=3}^{6} 2i =$

    a) 30
    b) 36
    c) 40
    d) 42

_____ **2.** True, False: $\displaystyle\sum_{i=3}^{7} i^2 = \sum_{j=3}^{7} j^2.$

_____ **3.** $3^3 + 4^3 + 5^3 =$

    a) $\displaystyle\sum_{i=1}^{3} i^3$

    b) $\displaystyle\sum_{i=3}^{5} 3i$

    c) $\displaystyle\sum_{i=1}^{5} i^3$

    d) $\displaystyle\sum_{i=3}^{5} i^3$

_____ **4.** $\displaystyle\sum_{i=1}^{n} (4i^2 + i) =$

    a) $\displaystyle 4\sum_{i=1}^{n} (i^2 + i)$

    b) $\displaystyle 4\sum_{i=1}^{n} i^2 + \sum_{i=1}^{n} i$

    c) $\displaystyle \sum_{i=1}^{n} 5i^2$

    d) $\displaystyle 4\sum_{i=1}^{n} i^2 + 4\sum_{i=1}^{n} i$

Section 5.2

_____ **1.** The norm of the partition $[4, 5]$, $[5, 7]$, $[7, 10]$, $[10, 12]$ of the interval $[4, 12]$ is

    a) 1
    b) 2
    c) 3
    d) 8

_____ **2.** Find the sum of the areas of approximating rectangles for the area under $f(x) = 48 - x^2$, between $x = 1$ and $x = 5$. Use the partition $[1, 2]$, $[2, 4]$ $[4, 5]$ and the right endpoints of each subinterval for x_i^*.

    a) 131
    b) 99
    c) 15
    d) 192

_____ **3.** True, False:

In general, a better approximation to the area under the curve $y = f(x)$ between $x = a$ and $x = b$ is obtained by selection of partitions with smaller norms.

_____ **1.** Sometimes, Always, or Never:

$\int_a^b f(x)\, dx$ equals the area between $y = f(x)$, the x-axis, $x = a$, and $x = b$.

_____ **2.** Sometimes, Always, or Never:

If $f(x)$ is continuous on $[a, b]$, then $\int_a^b f(x)\, dx$ exists.

_____ **3.** Using $n = 4$ and midpoints for x_i^*, the Riemann sum for $\int_{-1}^7 x^2\, dx$ is

    a) $\dfrac{344}{3}$
    b) 72
    c) 168
    d) 112

_____ **4.** If A_1 and A_2 are the areas at the right then $\int_a^b f(x)\, dx =$

    a) $A_1 - A_2$
    b) $-A_1 + A_2$
    c) $A_1 + A_2$
    d) $|A_1 - A_2|$.

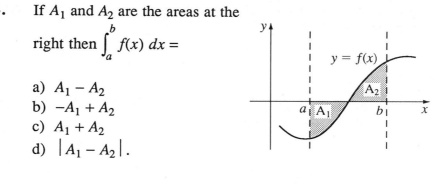

____ **1.** True, False: $\displaystyle\int_a^b f(x)\,g(x)\,dx = \left(\int_a^b f(x)\,dx\right)\left(\int_a^b g(x)\,dx\right).$

____ **2.** If $\displaystyle\int_1^3 f(x)\,dx = 10$ and $\displaystyle\int_1^3 g(x)\,dx = 6$, then $\displaystyle\int_1^3 (2f(x) - 3g(x))\,dx =$

    a) 2
    b) 4
    c) 18
    d) 38

____ **3.** If $f(x) \geq 5$ for all $x \in [2, 6]$ then $\displaystyle\int_2^6 f(x)\,dx \geq$ _____.
    (Choose the best answer.)

    a) 4
    b) 5
    c) 20
    d) 30

____ **4.** $\displaystyle\int_3^5 6\,dx =$

    a) 2
    b) 8
    c) 6
    d) 12

Section 5.5

___ **1.** If $h(x) = \int_3^x (8t^3 + 2t)\, dt$, then $h'(x) =$

    a) $8t^3 + 2t$
    b) $24x^2 + 2$
    c) $8x^3 + 2x - 3$
    d) $8x^3 + 2x$

___ **2.** $\int f(x)\, dx = F(x)$ means

    a) $f'(x) = F(x)$
    b) $f(x) = F'(x)$
    c) $f(x) = F(b) - F(a)$
    d) $f(x) = F(x) + C$

___ **3.** True, False: $\int_0^7 2\sqrt[3]{x}\, dx = 2\int_0^7 \sqrt[3]{x}\, dx$

___ **4.** Evaluate $\int_{-1}^1 (x^2 - x^3)\, dx$

    a) $\dfrac{3}{2}$

    b) $\dfrac{5}{6}$

    c) $\dfrac{1}{2}$

    d) $\dfrac{2}{3}$

___ **1.** Suppose $\int f(x) = F(x) + C$. Then $\int f(g(x))\, g'(x)\, dx =$

     a) $F(x)g(x) + C$
     b) $f'(g(x)) + C$
     c) $F(g'(x)) + C$
     d) $F(g(x)) + C$

___ **2.** What substitution should be made to evaluate $\int \sqrt{x^2 + 2x}\ (x + 1)\, dx$?

     a) $u = \sqrt{x^2 + 2x}$
     b) $u = x^2 + 2x$
     c) $u = x + 1$
     d) $u = \sqrt{x^2 + 2x}\ (x + 1)$

___ **3.** What substitution should be made to evaluate $\int \sin^2 x \cos x\, dx$?

     a) $u = \sin x$
     b) $u = \cos x$
     c) $u = \sin x \cos x$
     d) $u = \sin^2 x \cos x$

___ **4.** Evaluate $\int_0^1 (x^3 + 1)^2\, x^2\, dx$.

     a) $\dfrac{1}{9}$

     b) $\dfrac{1}{3}$

     c) $\dfrac{7}{9}$

     d) $\dfrac{8}{9}$

_____ **1.** $\int_1^7 \dfrac{1}{t}\, dt =$

    a) $\dfrac{1}{7}$

    b) $\ln 7$

    c) $\ln \dfrac{1}{7}$

    d) $\ln 6$

_____ **2.** For what number x is the shaded area equal to 2?

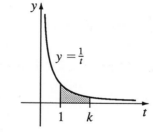

    a) $\ln 2$

    b) $\ln \dfrac{1}{2}$

    c) $2e$

    d) e^2

_____ 1.   A definite integral for the
area shaded at the right is:

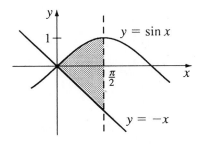

a) $\int_0^{\pi/2} [\sin x - x]\, dx$

b) $\int_0^{\pi/2} [\sin x + x]\, dx$

c) $\int_0^1 [\sin x - x]\, dx$

d) $\int_0^1 [\sin x + x]\, dx$

_____ 2.   For the area of the shaded region,
which integral form,

$\int \dots dx$ or $\int \dots dy$, should be used?

a) $\int \dots dx$

b) $\int \dots dy$

c) Either may be used with equal ease.

_____ 3. A definite integral for the area of the region bounded by
$y = 2 - x^2$ and $y = x^2$ is

a) $\int_{-1}^1 [2 - 2x^2]\, dx$

b) $\int_{-1}^1 [2x^2 - 2]\, dx$

c) $\int_0^2 [2 - 2x^2]\, dx$

d) $\int_0^2 [2x^2 - 2]\, dx$

Section 6.2

_____ **1.** Find a definite integral for the volume of a solid with a circular base of radius 4 in. such that the cross section of any slice perpendicular to a certain diameter in the base is a square.

a) $\int_{-4}^{4} (64 - 4x^2)\, dx$

b) $\int_{-4}^{4} 4x^2\, dx$

c) $\int_{-4}^{4} 4x^4\, dx$

d) $\int_{-4}^{4} (4 - x^2)^2\, dx$

cross section is a square

_____ **2.** An integral for the solid obtained by rotating the region at the right about the x-axis is

a) $\int_{0}^{1} \pi (x^4 - x^2)\, dx$

b) $\int_{0}^{1} \pi (x^2 - x^4)\, dx$

c) $\int_{0}^{1} \pi (x - x^2)\, dx$

d) $\int_{0}^{1} \pi (x^2 - x)\, dx$

_____ **3.** An integral for the solid obtained by rotating the region at the right about the y-axis is

a) $\int_{0}^{4} \pi y\, dy$

b) $\int_{0}^{2} \pi y^2\, dy$

c) $\int_{0}^{4} \pi \sqrt{y}\, dy$

d) $\int_{0}^{2} \pi \sqrt{y}\, dy$

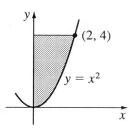

Section 6.3

_____ **1.** Which definite integral is the volume of the solid obtained by revolving the region at the right about the y-axis using cylindrical shells?

a) $\int_1^5 2\pi x^2 \, dx$

b) $\int_1^5 4\pi x^2 \, dx$

c) $\int_1^5 2\pi x \, dx$

d) $\int_2^{10} 4\pi x \, dx$

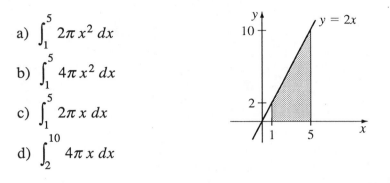

_____ **2.** Which definite integral is the volume of the solid obtained by revolving the region at the right about the x-axis using cylindrical shells?

a) $\int_1^{16} 2\pi y^3 \, dy$

b) $\int_1^{16} 2\pi \left(4 - \sqrt{x}\right) dy$

c) $\int_1^4 2\pi y^2 \, dy$

d) $\int_1^4 2\pi y^3 \, dy$

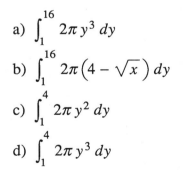
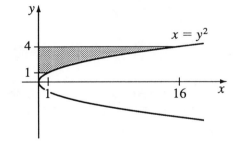

_____ **1.** How much work is done in lifting a 60 pound child 3 feet in the air?

    a) 20 ft-lb
    b) $20g$ ft-lb
    c) 180 ft-lb
    d) $180g$ ft-lb

_____ **2.** A particle moves along an x-axis from 2 m to 3 m pushed by a force of x^2 newtons. The integral that determines the amount of work done is

    a) $\displaystyle\int_2^3 g\, x^3\, dx$

    b) $\displaystyle\int_2^3 x\, dx$

    c) $\displaystyle\int_2^3 2\pi\, x^2\, dx$

    d) $\displaystyle\int_2^3 x^2\, dx$

_____ **1.** The average value of $f(x) = 3x^2 + 1$ on the interval $[2, 4]$ is

      a) 29
      b) 66
      c) 58
      d) 36

_____ **2.** Which point x_1, x_2, x_3, or x_4 on the graph is the best choice to serve as the point guaranteed by the Mean Value Theorem for Integrals?

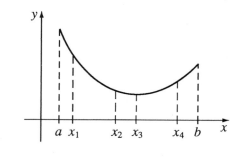

      a) x_1
      b) x_2
      c) x_3
      d) x_4

_____ **1.** Find the correct u and dv for integration by parts of $\int x^2 \cos 2x \, dx$.

    a) $u = x^2$, $dv = \cos 2x \, dx$
    b) $u = \cos 2x$, $dv = x^2 \, dx$
    c) $u = x^2$, $dv = 2x \, dx$
    d) $u = x \cos 2x$, $dv = x \, dx$

_____ **2.** Evaluate $\int 16 \, t^3 \ln t \, dt$.

    a) $4t^4 + C$
    b) $4t^3 \ln t - 16t^3 + C$
    c) $4t^4 - t^3 \ln t + C$
    d) $4t^4 \ln t - t^4 + C$

Section 7.2

_____ **1.** By the methods of trigonometric integrals, $\int \sin^3 x \cos^2 x \, dx$ should be rewritten as

a) $\int \sin^3 x \, (1 - \sin^2 x) \, dx$

b) $\int (1 - \cos^2 x)^{3/2} \cos^2 x \, dx$

c) $\int (\sin^3 x) \, (\cos x) \cos x \, dx$

d) $\int (1 - \cos^2 x) \cos^2 x \, \sin x \, dx$

_____ **2.** Using $\tan^2 x = \sec^2 x - 1$, $\int \tan^3 x \, dx$

a) $\dfrac{3}{2} \ln |\sec^2 x - 1| + C$

b) $\dfrac{1}{2} \tan^2 x - \ln |\sec x| + C$

c) $\dfrac{1}{2} \tan^2 x - \dfrac{1}{4} \sec^2 x + C$

d) $\dfrac{1}{4} \tan^2 x + \ln |\sec x| + C$

____ **1.** What trigonometric substitution should be made for $\int \dfrac{x^2}{\sqrt{x^2 - 25}}\, dx$?

    a) $x = 5 \sin \theta$
    b) $x = 5 \tan \theta$
    c) $x = 5 \sec \theta$
    d) No such substitution will evaluate the integral.

____ **2.** Rewrite $\int \dfrac{\sqrt{10 - x^2}}{x}\, dx$ using a trigonometric substitution.

    a) $\int \dfrac{\cos \theta}{\sqrt{10}\, \sin \theta}\, d\theta$

    b) $\int \dfrac{1}{\sqrt{10}}\, \sin \theta \cos^2\theta\, d\theta$

    c) $\int \dfrac{\sqrt{10}\, \cos^2\theta}{\sin \theta}\, d\theta$

    d) $\int \sqrt{10}\, \sin \theta \cos \theta\, d\theta$

_____ **1.** What is the partial fraction form of $\dfrac{2x + 3}{(x - 2)(x + 2)}$?

a) $\dfrac{Ax}{x - 2} + \dfrac{B}{x + 2}$

b) $\dfrac{A}{x - 2} + \dfrac{B}{x + 2}$

c) $\dfrac{Ax}{x - 2} + \dfrac{Bx}{x + 2}$

d) $\dfrac{Ax + B}{x - 2} + \dfrac{Cx + D}{x + 2}$

_____ **2.** What is the partial fraction form of $\dfrac{2x + 3}{(x^2 + 1)^2}$?

a) $\dfrac{Ax + B}{(x^2 + 1)^2}$

b) $\dfrac{A}{x^2 + 1} + \dfrac{B}{(x^2 + 1)^2}$

c) $\dfrac{A}{x^2 + 1} + \dfrac{Bx + C}{(x^2 + 1)^2}$

d) $\dfrac{Ax + B}{x^2 + 1} + \dfrac{Cx + D}{(x^2 + 1)^2}$

_____ **3.** Evaluate $\displaystyle\int \dfrac{2}{(x + 1)(x + 2)}\, dx$, using partial fractions.

a) $2 \ln |x + 1| - 2 \ln |x + 2|$
b) $\ln |x + 1| + 3 \ln |x + 2|$
c) $3 \ln |x + 1| - \ln |x + 2|$
d) $\ln |x + 1| + \ln |x + 2|$

Section 7.5

_____ **1.** What rationalizing substitution should be made for $\int \dfrac{\sqrt[3]{x} + 2}{\sqrt[3]{x} + 1} \, dx$?

    a) $u = x$
    b) $u = \sqrt[3]{x}$
    c) $u = \sqrt[3]{x} + 2$
    d) $u = \sqrt[3]{x} + 1$

_____ **2.** What rationalizing substitution should be made for $\int \dfrac{1}{\sqrt{x} + \sqrt[5]{x}} \, dx$?

    a) $u = x$
    b) $u = \sqrt{x}$
    c) $u = \sqrt[5]{x}$
    d) $u = \sqrt[10]{x}$

_____ **3.** What integral results from the substitution $t = \tan \dfrac{x}{2}$ in $\int \dfrac{\sin x}{1 - \cos x} \, dx$?

    a) $\int \dfrac{2}{1 + t^2} \, dt$

    b) $\int \dfrac{2t^2}{1 - t^2} \, dt$

    c) $\int \dfrac{4t^4}{1 + t^2} \, dt$

    d) $\int \dfrac{2t^2}{(1 + t^2)^2} \, dt$

_____ **1.** True, False:

$\int x^2 \sqrt{x^2 - 4} \ dx$ is evaluated using integration by parts.

_____ **2.** True, False:

A straight (simple) substitution may be used to evaluate $\int \dfrac{2 + \ln x}{x} \ dx$.

_____ **3.** True, False:

Partial fractions may be used for $\int \dfrac{1}{\sqrt{x + 2} \ \sqrt{x + 3}} \ dx$.

_____ **4.** True, False:

$\int \dfrac{\sqrt{x^2 + 1}}{x^3} \ dx$ may be solved using a trigonometric substitution.

_____ **5.** Sometimes, Always, or Never:

The antiderivative of an elementary function is elementary.

_____ **1.** What is the number of the integral formula in your text's Table of Integrals that may be used to evaluate $\int \dfrac{dx}{x \sqrt{2 - x^2}}$?

    a) #35
    b) #36
    c) #18
    d) #24

_____ **2.** What is the number of the integral formula in your text's Table of Integrals that may be used to evaluate $\int \dfrac{dx}{(x^2 + 2x + 5)^{3/2}}$?

    a) #37
    b) #38
    c) #29
    d) #46

_____ **1.** If $f(x)$ is integrable and concave upward on $[a, b]$, then an estimate
of $\int_a^b f(x)\,dx$ using the Trapezoidal Rule will always be

a) too large
b) too small
c) exact
d) within $\dfrac{b-a}{12n}$ of the exact value

_____ **2.** An estimate of $\int_{-1}^{3} x^4\,dx$ using Simpson's Rule with $n = 4$ gives

a) $\dfrac{242}{5}$

b) $\dfrac{148}{3}$

c) $\dfrac{152}{3}$

d) $\dfrac{244}{5}$

_____ **3.** Using $\dfrac{M(b-a)^3}{24\,n^2}$ find the maximum error in estimating $\int_{-1}^{3} x^4\,dx$
with the Midpoint Rule and 8 subintervals.

a) $\dfrac{1}{2}$

b) 9

c) $\dfrac{9}{2}$

d) $\dfrac{9}{8}$

_____ **1.** True, False:

$\int_{-2}^{3} |x|\, dx$ is an improper integral.

_____ **2.** By definition the improper integral $\int_{1}^{e} \dfrac{1}{x \ln x}\, dx =$

a) $\displaystyle\lim_{t \to 1^{+}} \int_{t}^{e} \dfrac{1}{x \ln x}\, dx$

b) $\displaystyle\lim_{t \to e^{-}} \int_{1}^{t} \dfrac{1}{x \ln x}\, dx$

c) $\displaystyle\int_{1}^{2} \dfrac{1}{x \ln x}\, dx + \int_{2}^{e} \dfrac{1}{x \ln x}\, dx$

d) It is not an improper integral.

_____ **3.** $\int_{0}^{2} \dfrac{x - 2}{\sqrt{4x - x^2}}\, dx =$

a) 0

b) 2

c) −2

d) does not exist

_____ **4.** Suppose we know $\int_{1}^{\infty} f(x)\, dx$ diverges and that $f(x) \geq g(x) \geq 0$ for

all $x \geq 1$. What conclusion can be made about $\int_{1}^{\infty} g(x)\, dx$?

a) it converges to the value of $\int_{1}^{\infty} f(x)\, dx$

b) it converges, but we don't know to what value.

c) it diverges

d) no conclusion (it could diverge or converge)

_____ **1.** A differential equation is separable if it can be written in the form

    a) $\dfrac{dy}{dx} = f(x) + g(y)$

    b) $\dfrac{dy}{dx} = f(g(x, y))$

    c) $\dfrac{dy}{dx} = \dfrac{f(x)}{g(y)}$

    d) $\dfrac{dy}{dx} = f(x)^{g(y)}$

_____ **2.** $(y''')^2 + 6xy - 2y' + 7x = 0$ has order

    a) 1
    b) 2
    c) 3
    d) 4

_____ **3.** The general solution to $\dfrac{dy}{dx} = \dfrac{1}{xy}$ (for $x, y > 0$) is

    a) $y = \ln x + C$
    b) $y = \sqrt{\ln x + C}$
    c) $y = \ln(\ln x + C)$
    d) $y = \sqrt{\ln x^2 + C}$

_____ **4.** The solution to $y' = -y^2$ with $y(1) = \dfrac{1}{3}$ is

    a) $y = \dfrac{1}{x + 2}$

    b) $y = \ln x + 3$

    c) $y = \dfrac{1}{2x^2} - \dfrac{1}{6}$

    d) $y = \dfrac{1}{2x} - \dfrac{1}{6}$

_____ **1.** A definite integral for the length of $y = x^3$, $1 \le x \le 2$ is

a) $\displaystyle\int_1^2 \sqrt{1 + x^3}\ dx$

b) $\displaystyle\int_1^2 \sqrt{1 + x^6}\ dx$

c) $\displaystyle\int_1^2 \sqrt{1 + 9x^4}\ dx$

d) $\displaystyle\int_1^2 \sqrt{1 + 3x^2}\ dx$

_____ **2.** The arc length function for $y = x^2$, $1 \le x \le 3$ is $s(x) =$

a) $\displaystyle\int_1^x \sqrt{1 + 4x^2}\ dx$

b) $\displaystyle\int_1^x \sqrt{1 + 4t^2}\ dt$

c) $\displaystyle\int_1^x \sqrt{1 + t^4}\ dt$

d) $\displaystyle\int_1^x \sqrt{1 + 4t^4}\ dt$

_____ **1.** An integral for the surface area obtained by rotating $y = \sin x + \cos x$, $0 \le x \le \pi/2$, about the x-axis is

a) $\displaystyle\int_0^{\pi/2} 2\pi (\sin x + \cos x) \sqrt{2 - 2\sin x \cos x}$

b) $\displaystyle\int_0^{\pi/2} 2\pi (\sin x + \cos x) \sqrt{1 + \sin^2 x - \cos^2 x}$

c) $\displaystyle\int_0^{\pi/2} 2\pi (\cos x - \sin x) \sqrt{1 + \sin^2 x - \cos^2 x}$

d) $\displaystyle\int_0^{\pi/2} 2\pi (\cos x - \sin x) \sqrt{2 - 2\sin x \cos x}$

_____ **2.** A hollow cylinder of radius 3 in. and height 4 in. is a surface of revolution. Its area may be expressed as

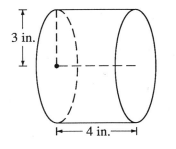

3 in.

4 in.

a) $\displaystyle\int_0^3 2\pi (4)\sqrt{1 + 0} \; dx$

b) $\displaystyle\int_0^3 2\pi (3)\sqrt{1 + 0} \; dx$

c) $\displaystyle\int_0^4 2\pi (3)\sqrt{1 + 0} \; dx$

d) $\displaystyle\int_0^4 2\pi (4)\sqrt{1 + 0} \; dx$

Section 8.4

_____ **1.** The x-coordinate of the center of mass of the region bounded by $y = \dfrac{1}{x}$, $x = 1$, $x = 2$, $y = 0$ is

    a) $\ln 2$

    b) $\dfrac{1}{\ln 2}$

    c) 1

    d) $2 \ln 2$

_____ **2.** The y-coordinate of the center of mass of the region bounded by $y = \sqrt{x - 1}$, $x = 1$, $x = 5$, $y = 0$ is

    a) $\dfrac{3}{14}$

    b) $\dfrac{6}{7}$

    c) $\dfrac{14}{3}$

    d) 4

_____ **1.** Find a definite integral for the hydrostatic force exerted by a liquid with density 800 kg/m³ on the semicircular plate in the center of the figure.

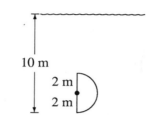

a) $\int_{-2}^{2} 800 \, (9.8) \, [8 - y]\left[2\sqrt{4 - y^2}\right] dy$

b) $\int_{-2}^{2} 800 \, (9.8) \, [8 - y] \sqrt{4 - y^2} \; dy$

c) $\int_{-2}^{2} 800 \, (9.8) \, [10 - y]\left[2\sqrt{4 - y^2}\right] dy$

d) $\int_{-2}^{2} 800 \, (9.8) \, [10 - y]\sqrt{4 - y^2} \; dy$

Section 8.6

_____ **1.** True, False:

A definite integral is used in an application to compute the total amount of a varying quantity.

_____ **2.** Find a definite integral for the consumer surplus if 180 units are available and the demand function is $p(x) = 3000 - 10x$.

a) $\int_0^{180} (3000 - 10x)\, dx$

b) $\int_0^{180} (1800 - 10x)\, dx$

c) $\int_0^{180} (1200 - 10x)\, dx$

d) $\int_0^{180} (180 - 10x)\, dx$

_____ **3.** Find a definite integral for the present value of an income stream of $5000 per year starting next year for 4 years if interest is compounded continuously at 8%.

a) $\int_1^5 5000\, e^{-.08t}\, dt$

b) $\int_1^5 5000\, t\, e^{-.08t}\, dt$

c) $\int_0^4 5000\, e^{-.08t}\, dt$

d) $\int_0^4 5000\, t\, e^{-.08t}\, dt$

_____ **1.** The graph of $x = 2 + 3t$, $y = 4 - t$ is a

    a) circle
    b) ellipse
    c) line
    d) parabola

_____ **2.** The best graph of $x = \cos t$, $y = \sin^2 t$ is

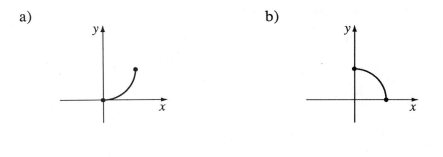

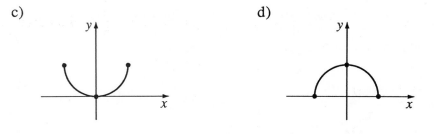

_____ **3.** Elimination of the parameter in $x = 2t^{3/2}$, $y = t^{2/3}$ gives

    a) $x^4 = 16y^9$
    b) $16x^4 = y^4$
    c) $x^3 = 8y^4$
    d) $8x^3 = y^4$

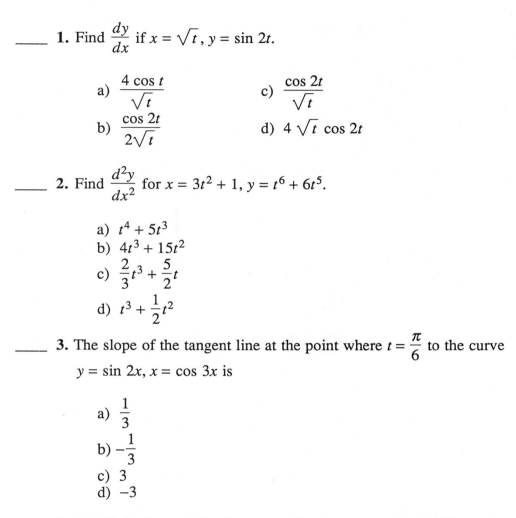

_____ **1.** Find $\dfrac{dy}{dx}$ if $x = \sqrt{t}$, $y = \sin 2t$.

     a) $\dfrac{4 \cos t}{\sqrt{t}}$ c) $\dfrac{\cos 2t}{\sqrt{t}}$

     b) $\dfrac{\cos 2t}{2\sqrt{t}}$ d) $4 \sqrt{t} \cos 2t$

_____ **2.** Find $\dfrac{d^2y}{dx^2}$ for $x = 3t^2 + 1$, $y = t^6 + 6t^5$.

     a) $t^4 + 5t^3$
     b) $4t^3 + 15t^2$
     c) $\dfrac{2}{3}t^3 + \dfrac{5}{2}t$
     d) $t^3 + \dfrac{1}{2}t^2$

_____ **3.** The slope of the tangent line at the point where $t = \dfrac{\pi}{6}$ to the curve $y = \sin 2x$, $x = \cos 3x$ is

     a) $\dfrac{1}{3}$

     b) $-\dfrac{1}{3}$

     c) 3
     d) -3

_____ **4.** A definite integral for the area under the curve described by $x = t^2 + 1$, $y = 2t$, $0 \le t \le 1$ is

     a) $\displaystyle\int_0^1 (2t^3 + 2t)\,dt$ c) $\displaystyle\int_0^1 (2t^2 + 2)\,dt$

     b) $\displaystyle\int_0^1 4t^2\,dt$ d) $\displaystyle\int_0^1 4t\,dt$

_____ **1.** The length of the curve given by $x = 3t^2 + 2$, $y = 2t^3$, $t \in [0, 1]$ is

    a) $4\sqrt{2} - 2$

    b) $8\sqrt{2} - 1$

    c) $\frac{2}{3}\left(2\sqrt{2} - 1\right)$

    d) $\sqrt{2} - 1$

_____ **2.** Find a definite integral for the area of the surface of revolution about the x-axis obtained by rotating the curve $y = t^2$, $x = 1 + 3t$, $0 \leq t \leq 2$.

    a) $\displaystyle\int_0^2 2\pi\, t^2 \sqrt{t^4 + 9t^2 + 6t + 1}\ dt$

    b) $\displaystyle\int_0^2 2\pi\, t^2 \sqrt{4t^2 + 9}\ dt$

    c) $\displaystyle\int_0^2 2\pi\, (2t)\sqrt{t^4 + 9t^2 + 6t + 1}\ dt$

    d) $\displaystyle\int_0^2 2\pi\, (2t)\sqrt{4t^2 + 9}\ dt$

_____ **1.** The polar coordinates of the point plotted at the right are:

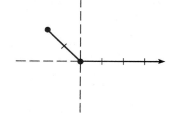

    a) $(-2, \pi/4)$
    b) $(-2, 3\pi/4)$
    c) $(2, -\pi/4)$
    d) $(2, 3\pi/4)$

_____ **2.** Polar coordinates of the point with rectangular coordinates $(5, 5)$ are

    a) $(25, 0)$
    b) $(5, \pi/4)$
    c) $(5\sqrt{5}, \pi/4)$
    d) $(50, -\pi/4)$

_____ **3.** Rectangular coordinates of the point with polar coordinates $(-1, 3\pi/2)$ are

    a) $(-1, 0)$
    b) $(0, 1)$
    c) $(0, -1)$
    d) $(1, 0)$

_____ **4.** The graph of $\theta = 2$ in polar coordinates is a

    a) circle
    b) line
    c) spiral
    d) 3-leaved rose

_____ **5.** The slope of the tangent line to $r = \cos \theta$ at $\theta = \pi/3$ is

    a) $\sqrt{3}$
    b) $\dfrac{1}{\sqrt{3}}$
    c) $-\sqrt{3}$
    d) $-\dfrac{1}{\sqrt{3}}$

_____ **1.** The area of the region bounded by $\theta = \pi/3$, $\theta = \pi/4$, and $r = \sec\theta$ is

   a) $\dfrac{1}{2}\left(\sqrt{3} - 1\right)$

   b) $\sqrt{3}$

   c) $2\left(\sqrt{3} - 1\right)$

   d) $2\sqrt{3}$

_____ **2.** The area of the shaded region is given by

   a) $\displaystyle\int_0^{\pi/2} \sin 3\theta \, d\theta$

   b) $\displaystyle\int_0^{\pi/2} 2\sin^2 3\theta \, d\theta$

   c) $\displaystyle\int_0^{\pi/3} \sin 3\theta \, d\theta$

   d) $\displaystyle\int_0^{\pi/3} 2\sin^2 3\theta \, d\theta$

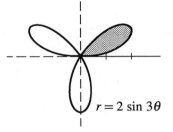
$r = 2\sin 3\theta$

_____ **3.** The length of the arc $r = e^{\theta}$ for $0 \le \theta \le \pi$ is

   a) $e^{\pi} - 1$

   b) $2(e^{\pi} - 1)$

   c) $\sqrt{2}\,(e^{\pi} - 1)$

   d) $2\sqrt{2}\,(e^{\pi} - 1)$

_____ **1.** The parabola at the right has equation

    a) $y^2 = -12x$
    b) $y^2 = -3x$
    c) $x^2 = -3y$
    d) $x^2 = -12y$

_____ **2.** The ellipse at the right has equation

    a) $\dfrac{x^2}{4} + \dfrac{y^2}{9} = 1$
    c) $\dfrac{x^2}{9} + \dfrac{y^2}{4} = 1$

    b) $\dfrac{x^2}{5} + \dfrac{y^2}{9} = 1$
    d) $\dfrac{x^2}{9} + \dfrac{y^2}{5} = 1$

_____ **3.** The hyperbola at the right has equation

    a) $\dfrac{x^2}{49} - \dfrac{y^2}{4} = 1$
    c) $\dfrac{y^2}{49} - \dfrac{x^2}{4} = 1$

    b) $\dfrac{x^2}{4} - \dfrac{y^2}{49} = 1$
    d) $\dfrac{y^2}{4} - \dfrac{x^2}{49} = 1$

_____ **4.** The set of all points P such that the sum of the distances from P to two fixed points is a constant is a(n)

    a) parabola
    b) ellipse
    c) hyperbola
    d) not a conic section

_____ **5.** The equation $4x^2 + 16x = 9y^2 + 81y$ represents a(n)

    a) parabola
    b) ellipse
    c) hyperbola
    d) not a conic section

_____ **1.** The figure at the right shows one point
P on a conic and the distances of P to
the focus and the directrix of a conic.
What type of conic is it?

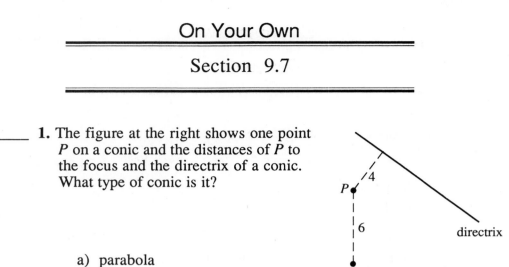

    a) parabola
    b) ellipse
    c) hyperbola
    d) Not enough information is provided to answer the question.

_____ **2.** The polar equation of the conic with eccentricity 3 and directrix
$x = -7$ is

    a) $r = \dfrac{21}{1 + 3 \cos \theta}$

    b) $r = \dfrac{21}{1 - 3 \cos \theta}$

    c) $r = \dfrac{21}{1 + 3 \sin \theta}$

    d) $r = \dfrac{21}{1 - 3 \sin \theta}$

_____ **3.** The directrix of the conic given by $r = \dfrac{6}{2 + 10 \sin \theta}$ is

    a) $x = \dfrac{5}{3}$

    b) $x = \dfrac{3}{5}$

    c) $y = \dfrac{5}{3}$

    d) $y = \dfrac{3}{5}$

____ **1.** $\lim\limits_{n \to \infty} \dfrac{n^2 + 3n}{2n^2 + n + 1} =$

     a) 0

     b) $\dfrac{1}{2}$

     c) 1

     d) ∞

____ **2.** Sometimes, Always, or Never:

If $a_n \geq b_n \geq 0$ and $\{b_n\}$ diverges, then $\{a_n\}$ diverges.

____ **3.** Sometimes, Always, or Never:

If $a_n \leq b_n \leq c_n$ and both $\{a_n\}$ and $\{c_n\}$ converge, then $\{b_n\}$ converges.

____ **4.** Sometimes, Always, or Never:

If $\{a_n\}$ is increasing and bounded above, then $\{a_n\}$ converges.

____ **5.** True, False: $\quad a_n = \dfrac{(-1)^n}{n^2}$ is monotonic.

____ **6.** $\lim\limits_{n \to \infty} \dfrac{\arctan n}{2} =$

     a) $\pi/4$

     b) $\pi/2$

     c) π

     d) does not exist

On Your Own

Section 10.2

_____ **1.** True, False:

$$\sum_{n=1}^{\infty} a_n \text{ converges if } \lim_{n \to \infty} a_n = 0.$$

_____ **2.** True, False:

If $\sum_{n=1}^{\infty} a_n$ converges and $\sum_{n=1}^{\infty} b_n$ converges, then $\sum_{n=1}^{\infty} a_n - b_n$ converges.

_____ **3.** The harmonic series is

    a) $1 + 2 + 3 + 4 + \ldots$
    b) $1 + \frac{1}{2} + \frac{1}{3} + \frac{1}{4} + \ldots$
    c) $1 + \frac{1}{2} + \frac{1}{4} + \frac{1}{8} + \ldots$
    d) $1 - \frac{1}{2} + \frac{1}{4} - \frac{1}{8} + \ldots$

_____ **4.** True, False:

$$\sum_{n=1}^{\infty} \frac{n}{n+1} \text{ converges.}$$

_____ **5.** $\sum_{n=1}^{\infty} 2 \left(\frac{1}{4}\right)^n$ converges to

    a) $\dfrac{9}{4}$
    b) 2
    c) $\dfrac{8}{3}$
    d) The series diverges.

_____ **6.** True, False: $-3 + 1 - \frac{1}{3} + \frac{1}{9} - \frac{1}{27} + \ldots$ is a geometric series.

_____ **1.** For what values of p does the series $\displaystyle\sum_{n=1}^{\infty} \frac{1}{(n^2)^p}$ converge?

a) $p > -\dfrac{1}{2}$

b) $p < -\dfrac{1}{2}$

c) $p > \dfrac{1}{2}$

d) $p < \dfrac{1}{2}$

_____ **2.** True, False:

If $f(n) = a_n$ for all n, and $f(x)$ is continuous, decreasing, and $\displaystyle\int_{1}^{\infty} f(x)\, dx = M$,

then $\displaystyle\sum_{n=1}^{\infty} a_n = M$.

_____ **3.** Does $\displaystyle\sum_{n=1}^{\infty} \frac{1}{n^{2/3}}$ converge?

_____ **4.** Does $\displaystyle\sum_{n=1}^{\infty} \frac{n+2}{(n^2 + 4n + 1)^2}$ converge?

Section 10.4

_____ **1.** Sometimes, Always, or Never:

If $0 \leq a_n \leq b_n$ for all n and $\displaystyle\sum_{n=1}^{\infty} a_n$ diverges, then $\displaystyle\sum_{n=1}^{\infty} b_n$ converges.

_____ **2.** Does $\displaystyle\sum_{n=1}^{\infty} \frac{n+1}{n^3}$ converge?

_____ **3.** Does $\displaystyle\sum_{n=1}^{\infty} \frac{\sqrt{n} + \sqrt[3]{n}}{n^{2/3} + n^{3/2} + 1}$ converge?

_____ **4.** Does $\displaystyle\sum_{n=1}^{\infty} \frac{\cos^2(2^n)}{2^n}$ converge?

_____ **1.** Does $\displaystyle\sum_{n=1}^{\infty} \frac{(-1)^{n+1}\ln n}{n^2}$ converge?

_____ **2.** Does $\displaystyle\sum_{n=1}^{\infty} \frac{(-1)^n}{\sqrt[4]{n+1}}$ converge?

_____ **3.** For what value of n is the nth partial sum within .001 of the value of $\displaystyle\sum_{n=1}^{\infty} \frac{(-1)^n}{2^n}$?

    a) $n = 4$
    b) $n = 6$
    c) $n = 8$
    d) $n = 10$

_____ **4.** True, False:

The Alternating Series Test may be applied to determine the convergence of $\displaystyle\sum_{n=1}^{\infty} \frac{1 + (-1)^n}{2n^2}$.

_____ **1.** True, False:

If $\displaystyle\sum_{n=1}^{\infty} a_n$ converges absolutely, then it converges conditionally.

_____ **2.** True, False:

If $\displaystyle\lim_{n\to\infty} \left| \frac{a_n}{a_{n+1}} \right| = 3$, then $\displaystyle\sum_{n=1}^{\infty} a_n$ converges absolutely.

_____ **3.** True, False:

Every series must either converge absolutely, converge conditionally, or diverge.

_____ **4.** The series $\displaystyle\sum_{n=1}^{\infty} 2^{-n} n!$

    a) diverges
    b) converges absolutely
    c) converges conditionally
    d) converges, but not absolutely nor conditionally.

_____ **5.** The series $\displaystyle\sum_{n=1}^{\infty} \frac{(-5)^{n+1}}{n^n}$

    a) diverges
    b) converges absolutely
    c) converges conditionally
    d) converges, but not absolutely nor conditionally.

_____ **1.** True, False: $\displaystyle\sum_{n=2}^{\infty} \frac{1}{(\ln n)^n}$ converges.

_____ **2.** True, False: $\displaystyle\sum_{n=1}^{\infty} \frac{6}{7n+8}$ converges.

_____ **3.** True, False: $\displaystyle\sum_{n=1}^{\infty} \frac{(-1)^n \sqrt{n}}{n+3}$ converges.

_____ **4.** True, False: $\displaystyle\sum_{n=1}^{\infty} \frac{e^n}{n!}$ converges.

Section 10.8

_____ **1.** Sometimes, Always, or Never:

The interval of convergence of a power series $\sum_{n=0}^{\infty} a_n (x - c)^n$ is an open interval $(c - R, c + R)$ (where when $R = 0$ we mean $\{0\}$ and when $R = \infty$ we mean $(-\infty, \infty)$).

_____ **2.** True, False:

If a number p is in the interval of convergence of $\sum_{n=0}^{\infty} a_n x^n$ then so is the number $\frac{p}{2}$.

_____ **3.** For $f(x) = \sum_{n=0}^{\infty} \frac{(x - 1)^n}{3^n}$, $f(3) =$

a) 0
b) 2
c) 3
d) $f(3)$ does not exist.

_____ **4.** The interval of convergence of $\sum_{n=1}^{\infty} \frac{x^n}{\sqrt{n}}$ is

a) $[-1, 1]$
b) $[-1, 1)$
c) $(-1, 1]$
d) $(-1, 1)$

_____ **5.** The radius of convergence of $\sum_{n=0}^{\infty} \frac{n(x - 5)^n}{3^n}$ is

a) $\frac{1}{3}$
b) 1
c) 3
d) ∞

_____ **1.** For $f(x) = \displaystyle\sum_{n=0}^{\infty} \frac{x^{2n}}{n}$, $f'(x) =$

  a) $\displaystyle\sum_{n=1}^{\infty} x^{2n-1}$

  b) $\displaystyle\sum_{n=1}^{\infty} 2x^{2n-1}$

  c) $\displaystyle\sum_{n=1}^{\infty} 2^n x^{2n-1}$

  d) $\displaystyle\sum_{n=1}^{\infty} (2n-1) x^{2n-1}$

_____ **2.** Given the Taylor series $e^x = \displaystyle\sum_{n=0}^{\infty} \frac{x^n}{n!}$, a Taylor series for $e^{x/2}$ is

  a) $\displaystyle\sum_{n=0}^{\infty} \frac{2^n x^n}{n!}$

  b) $\displaystyle\sum_{n=0}^{\infty} \frac{2x^n}{n!}$

  c) $\displaystyle\sum_{n=0}^{\infty} \frac{x^n}{2^n n!}$

  d) $\displaystyle\sum_{n=0}^{\infty} \frac{x^n}{2 n!}$

_____ **3.** The nth term in the Taylor series about $x = 1$ for $f(x) = x^{-2}$ is

  a) $(-1)^{n+1}(n+1)! \, (x-1)^n$
  b) $(-1)^n \, n! \, (x-1)^n$
  c) $(-1)^{n+1}(n+1) \, (x-1)^n$
  d) $(-1)^n \, n \, (x-1)^n$

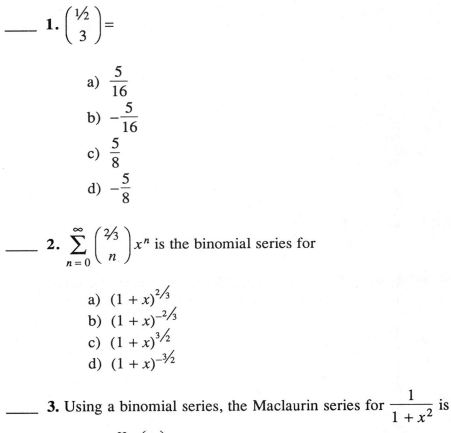

_____ **1.** $\dbinom{\frac{1}{2}}{3} =$

    a) $\dfrac{5}{16}$

    b) $-\dfrac{5}{16}$

    c) $\dfrac{5}{8}$

    d) $-\dfrac{5}{8}$

_____ **2.** $\displaystyle\sum_{n=0}^{\infty} \dbinom{\frac{2}{3}}{n} x^n$ is the binomial series for

    a) $(1+x)^{\frac{2}{3}}$
    b) $(1+x)^{-\frac{2}{3}}$
    c) $(1+x)^{\frac{3}{2}}$
    d) $(1+x)^{-\frac{3}{2}}$

_____ **3.** Using a binomial series, the Maclaurin series for $\dfrac{1}{1+x^2}$ is

    a) $\displaystyle\sum_{n=0}^{\infty} \dbinom{1}{n} x^n$

    b) $\displaystyle\sum_{n=0}^{\infty} \dbinom{-1}{n} x^n$

    c) $\displaystyle\sum_{n=0}^{\infty} \dbinom{1}{n} x^{2n}$

    d) $\displaystyle\sum_{n=0}^{\infty} \dbinom{-1}{n} x^{2n}$

_____ **1.** The Taylor polynomial of degree 3 for $f(x) = x(\ln x - 1)$ about $x = 1$ is $T_3(x) =$

a) $-1 + \dfrac{x^2}{x} - \dfrac{x^3}{6}$

b) $-1 + x - \dfrac{x^2}{2} + \dfrac{x^3}{6}$

c) $-1 + x - x^2 + x^3$

d) $-1 - x + x^2 - x^3$

_____ **2.** True, False:

If $T_n(x)$ is the nth Taylor polynomial for $f(x)$ at $x = c$, then $T_n^{(k)}(c) = f^{(k)}(c)$ for $k = 0, 1, \ldots, n$.

_____ **3.** If the Maclaurin polynomial of degree 2 for $f(x) = e^x$ is used to approximate $e^{.2}$ then an estimate for the error with $0 < z < .2$ is

a) $\dfrac{1}{24} e^z$

b) $\dfrac{1}{6} e^z$

c) $\dfrac{1}{2} e^z$

d) e^z

Answers to On Your Own

Section RP 1

1. True
2. False
3. C
4. C
5. Sometimes
6. A

Section RP 2

1. D
2. False
3. A
4. B
5. D

Section RP 3

1. D
2. D
3. B
4. C

Section RP 4

1. True
2. False
3. B
4. True
5. B

Section RP 5

1. C
2. A
3. A
4. Sometimes
5. B

Section RP 6

1. False
2. False
3. D
4. True

Section 1.1

1. B
2. B
3. False

Section 1.2

1. B
2. A
3. A
4. D

Section 1.3

1. True
2. C
3. Sometimes
4. A
5. B
6. B

Section 1.4

1. True
2. B

Section 1.5

1. Sometimes
2. C
3. False
4. A
5. False

Section 1.6

1. Sometimes
2. C
3. True
4. D
5. True

Section 1.7

1. C
2. A
3. Sometimes
4. C

Section 1.8

1. True
2. False
3. B
4. True
5. C

Section 2.1

1. False
2. True
3. D
4. B
5. B
6. True

Section 2.2

1. D
2. C
3. False
4. D
5. D

Section 2.3

1. False
2. A
3. B

Section 2.4

1. B
2. A
3. D
4. C

Section 2.5

1. False
2. C
3. D
4. A

Section 2.6

1. False
2. B
3. A

Section 2.7

1. True
2. D
3. B
4. B
5. C

Section 2.8

1. A
2. C

Section 2.9

1. B
2. C
3. D
4. B

Section 2.10

1. Sometimes
2. A

Section 3.1

1. Sometimes
2. B
3. A
4. C

Section 3.2

1. Sometimes
2. C
3. C
4. B

Section 3.3

1. C
2. D
3. Always
4. D

Section 3.4

1. B
2. False
3. A
4. D
5. A

Section 3.5

1. C
2. B
3. D
4. C

Section 3.6

1. A
2. C

Section 3.7

1. B
2. D
3. B

Section 3.8

1. False
2. B
3. D
4. C

Section 3.9

1. B
2. A
3. B

Section 4.1

1. False
2. True
3. C
4. True
5. B
6. B

Section 4.2

1. C
2. No
3. A
4. False

Section 4.3

1. A
2. False
3. False
4. B

Section 4.4

1. False
2. B
3. C

Section 4.5

1. True
2. False
3. False
4. True
5. False
6. True
7. B
8. D

Section 4.6

1. B
2. C

Section 4.7

1. False
2. True
3. B
4. B
5. B

Section 4.8

1. False
2. D
3. A

Section 5.1

1. B
2. True
3. D
4. B

Section 5.2

1. C
2. A
3. True

Section 5.3

1. Sometimes
2. Always
3. D
4. B

Section 5.4

1. False
2. A
3. C
4. D

Section 5.5

1. D
2. B
3. True
4. D

Section 5.6

1. D
2. B
3. A
4. C

Section 5.7

1. B
2. D

Section 6.1

1. B
2. B
3. A

Section 6.2

1. A
2. B
3. A

Section 6.3

1. B
2. D

Section 6.4

1. C
2. D

Section 6.5

1. A
2. D

Section 7.1

1. A
2. D

Section 7.2

1. D
2. B

Section 7.3

1. C
2. C

Section 7.4

1. B
2. D
3. A

Section 7.5

1. B
2. D
3. A

Section 7.6

1. False
2. True
3. False
4. True
5. Sometimes

Section 7.7

1. A
2. C

Section 7.8

1. A
2. B
3. C

Section 7.9

1. False
2. A
3. C
4. D

Section 8.1

1. C
2. C
3. D
4. A

Section 8.2

1. C
2. B

Section 8.3

1. A
2. C

Section 8.4

1. B
2. B

Section 8.5

1. A

Section 8.6

1. True
2. B
3. A

Section 9.1

1. C
2. D
3. A

Section 9.2

1. D
2. C
3. B
4. B

Section 9.3

1. A
2. B

Section 9.4

1. D
2. C
3. B
4. B
5. B

Section 9.5

1. A
2. D
3. C

Section 9.6

1. A
2. B
3. B
4. B
5. C

Section 9.7

1. C
2. B
3. D

Section 10.1

1. B
2. Sometimes
3. Sometimes
4. Always
5. False
6. A

Section 10.2

1. False
2. True
3. B
4. False
5. C
6. True

Section 10.3

1. C
2. False
3. No
4. Yes

Section 10.4

1. Never
2. Yes
3. No
4. Yes

Section 10.5

1. Yes
2. Yes
3. D
4. False

Section 10.6

1. False
2. True
3. True
4. A
5. B

Section 10.7

1. True
2. False
3. True
4. True

Section 10.8

1. Sometimes
2. True
3. C
4. B
5. C

Section 10.9

1. B
2. C
3. C